Perspectives in Sustainable Equity Investing

Perspectives in Sustainable Equity Investing

Guillaume Coqueret

CRC Press
Taylor & Francis Group
Boca Raton   London   New York

CRC Press is an imprint of the
Taylor & Francis Group, an **informa** business
A CHAPMAN & HALL BOOK

First edition published 2022
by CRC Press
6000 Broken Sound Parkway NW, Suite 300, Boca Raton, FL 33487-2742

and by CRC Press
4 Park Square, Milton Park, Abingdon, Oxon, OX14 4RN

CRC Press is an imprint of Taylor & Francis Group, LLC

© 2022 Guillaume Coqueret

Library of Congress Cataloging-in-Publication Data
Names: Coqueret, Guillaume, author. Title: Perspectives in sustainable equity investing / Guillaume Coqueret. Description: 1 Edition.	Boca Raton : CRC Press, 2022.	Includes bibliographical references and index. Identifiers: LCCN 2021040991 (print)	LCCN 2021040992 (ebook)	ISBN 9781032071015 (hardback)	ISBN 9781032104188 (paperback)	ISBN 9781003215257 (ebook) Subjects: LCSH: Investments--Social aspects.	Investments--Environmental aspects.	Social responsibility of business. Classification: LCC HG4515.13 .C75 2022 (print)	LCC HG4515.13 (ebook) | DDC 332.6--dc23/eng/20211014 LC record available at https://lccn.loc.gov/2021040991 LC ebook record available at https://lccn.loc.gov/2021040992

ISBN: 978-1-032-07101-5 (hbk)
ISBN: 978-1-032-10418-8 (pbk)
ISBN: 978-1-003-21525-7 (ebk)

DOI: 10.1201/9781003215257

Publisher's note: This book has been prepared from camera-ready copy provided by the authors.

Access the Support Material: https://www.routledge.com/9781032071015

To Sacha and Noé

- may they grow and thrive in a breathable world -

Contents

Acknowledgments   ix

CHAPTER 1 ▪ Introduction   1

1.1	CONTENT AND TARGET AUDIENCE	1
1.2	FOREWORD	2
1.3	PREVIOUS SURVEYS AND META-STUDIES	5

CHAPTER 2 ▪ ESG data   11

2.1	OVERVIEW OF ESG ISSUES		12
	2.1.1	Traditional issues	12
	2.1.2	Missing data	17
	2.1.3	Sovereign metrics	17
2.2	RATING DISAGREEMENT		19
2.3	MEASUREMENT ISSUES AND STABILITY		21
2.4	GREENWASHING		22
2.5	GREEN FIRMS		25
2.6	AD-HOC SOLUTIONS		27
2.7	NEED FOR TRANSPARENT AND UNIFORM REPORTING		28

CHAPTER 3 ▪ Investors and SRI   31

3.1	INVESTOR PREFERENCES AND BELIEFS	31
3.2	IMPACT INVESTING	38
3.3	INVESTOR PRACTICES	43

CHAPTER 4 ▪ ESG investing and financial performance   45

4.1	TOY MODEL		45
	4.1.1	Theory: assets, agents, equilibrium	46
	4.1.2	Numerical example	49
4.2	SRI IMPROVES PERFORMANCE		51
4.3	SRI DOES NOT IMPACT PERFORMANCE		55

4.4	SRI IS FINANCIALLY DETRIMENTAL	57
4.5	IT DEPENDS...	60
4.6	CSR AND RISK	66
4.7	ESG AND OTHER FINANCIAL METRICS	68
4.8	EMPIRICAL ILLUSTRATION	70

CHAPTER 5 ▪ Quantitative portfolio construction with ESG data and criteria   **75**

5.1	SIMPLE PORTFOLIO CHOICE SOLUTIONS	75
5.2	IMPROVED MEAN-VARIANCE ALLOCATION	76
5.3	OTHER QUANTITATIVE TECHNIQUES	80
5.4	MISCELLANEOUS TIPS, METHODS AND OTHER INTEGRATION TECHNIQUES	81

CHAPTER 6 ▪ Climate change risk   **85**

6.1	UNCERTAIN DISCOUNTING	86
6.2	MEASUREMENT ISSUES	89
6.3	STRESS TESTS AND OTHER MEASURES	92
6.4	MICRO- AND MACRO-ECONOMIC IMPACTS	94
6.5	INVESTOR ATTENTION	96
6.6	POLICY	97

CHAPTER 7 ▪ Conclusion   **103**

Bibliography   **105**

Index   **193**

Acknowledgments

I thank Serge Houles, Arne Hassel, Christian Morgenstern and James Kelly for spurring me to take up this challenge. I am indebted to Alban Cousin for our many discussions, to Justina Lee for an incredible amount of suggestions and to Jean-Michel Maeso for giving me access to practitioner-orientated parts of the literature. I am also grateful to Alex Edmans for advising a new positioning of the original paper. Finally, this book could not have been written without the confidence and help of David Grubbs and Jessica Poile.

Guillaume Coqueret
EMLYON Business School,
23 avenue Guy de Collongue,
69130 Ecully, FRANCE.
E-mail: coqueret@em-lyon.com

CHAPTER 1

Introduction

1.1 CONTENT AND TARGET AUDIENCE

This short book is a large-scale literature review on the topic of sustainable equity investing. It covers more than 900 academic sources (research articles, overwhelmingly), grouped into six thematic chapters. As such, it can serve as reference for analysts and researchers who work on environmental, social or governance (ESG)-driven portfolios. The book assumes no prior knowledge of this field, but some parts, especially the last chapter, can be rather technical mathematically.

While the sheer number of references may seem exhaustive, the survey only scratches the surface of the topic because the amount of contributions is daunting. The best round estimate on the pace of the development of the literature is roughly two serious papers per day (based on one year of compiling research output). The intellectual production on this matter is such that surveying a narrow subfield thereof can take weeks. In addition, all facets of the issue are interconnected, so that it does not make any sense to consider them separately. Focusing on pure ESG integration without notions in climate change risk is ill-advised. The big picture matters.

Briefly, we outline the chapters of the book below. Chapter 2 introduces a large number of terms to make sure the reader familiarizes with the jargon and nomenclature. In Chapter 3, we cover the broad theme of *green investors*, including their beliefs, preferences, and practices, as well as the topic of impact investing. Chapter 4, which is the core of the book, is dedicated to the very complex relationship between ESG investing and financial performance. Technical details on how to integrate sustainable criteria in complex portfolio optimization are provided in Chapter 5. The subject of climate change (its measurement, impact, and how to tackle it) is treated in Chapter 6. Finally, an online Chapter[1] is dedicated to theoretical models related to ESG investing and to the push for sustainability more broadly in the economy.

The review is compact, meaning that we favored concision to in-depth treatment. The monograph is intended as a thematic compass, pointing towards some relevant directions. For many topics, the interested reader will have to satisfy his or her curiosity by examining the mentioned references.

While the bulk of the book is non-technical, it is written in an academic fashion and the density of references is significantly more pronounced, compared to monographs aimed at popularization. As such, prior knowledge of notions related to ESG investing can prove helpful, even if we cover many of them along the way. Consequently, the readers who will most benefit from this work are professionals who work in the field of corporate sustainability and/or in the money management industry. Students and scholars may also appreciate a very compact overview of most themes that relate to green equity investing.

1.2   FOREWORD

The output of academic research linked to the broad topic of *sustainability* is growing exponentially. The reason behind this dynamic is patent: the durable negligence of resources (both natural and human) is increasingly perceived as an unstoppable threat to economic, cultural and social activities. This rising awareness is spreading at many levels and through

---

[1]https:///www.routledge.com/9781032071015

diverse channels. In particular, investors are increasingly concerned with these issues and their preferences are shifting: environment, social and governance (ESG) preoccupations have become ubiquitous in the money management industry.[2] Concomitantly, firms, because they genuinely believe in sustainability, or because they want to attract capital, also bend some of their corporate policies toward green or social goals.

This trend stands in sharp contrast with Milton Friedman's 1970 claim that *"The Social Responsibility of Business is to Increase Its Profits"* (which is the title of a short article in the *New York Times* magazine - see Friedman (1970)).[3] The arguments in the paper are subject to interpretation (see Hart and Zingales (2017)) and must be contextualized (Austin (2020)), but it is undeniable that the discussions around corporate social responsibility (CSR) have since then spawned a rich and sometimes passionate debate.[4] The resulting and rapidly expanding literature, stemming both from scholars and practitioners,[5] is notoriously hard to survey. As Nath (2021) (p. 194) puts it: *"SRI evolved from having a simple definition to a more complex concept embodying heterogeneity in terminology, definitions, strategies and practices"*.

The organization of the book is intended to make it useful to researchers or practitioners (e.g., quantitative analysts, or portfolio managers) who seek references on a particular subfield of socially responsible investing (SRI) - or who want to make sure their own literature review is not missing important

---

[2]See, e.g., CNBC's report, The Rise of ESG Investing.

[3]Surprisingly, maximizing profitability may sometimes be beneficial from a social point of view - see Green and Roth (2020).

[4]Technically, it is hard to date the origins of socially responsible investing (SRI). We refer to Townsend (2020) for a historical perspective on the matter. Cyert and March (1963) is often cited as a foundational work on the topic. The authors advocate that firms should take into account all stakeholders and not only maximize value for the shareholder.

[5]Recently, the Portfolio Management Research umbrella even created a dedicated outlet: the *Journal of Impact and ESG Investing*. Meanwhile, the *Journal of Sustainable Finance & Investment* is in its tenth year, and its contributions are reviewed with a bibliometric lens in Alshater et al. (2021).

or relevant references. The survey is focused on the equity investment space, while other asset classes (e.g., green bonds or
sustainable real estate) are out of the scope of the present
compilation. One of its intentions is to leave room for recent contributions, hence it incorporates many working papers
(consequently, this implies heterogeneity in the quality of the
references). In Figure 1.1, we show the outlets which we cite
most often (by numbers of papers) as well as the dynamic in
publication dates. Whenever an unpublished manuscript could
be found on the Social Science Research Network (SSRN),
we picked SSRN as the journal affiliation by default, which
explains its prominence in the left panel. In Figure 1.2, we
exhibit the most frequent words in the titles of the references.

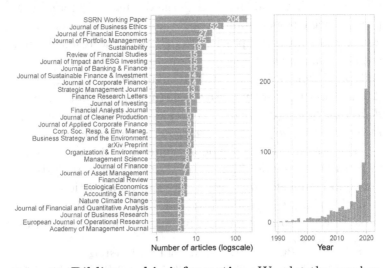

Figure 1.1 **Bibliographic information**. We plot the number
of articles per journal (left) and per year (right). Journals with
fewer than 5 articles are omitted and the dynamic before 1990
is not reported. The total number of references included in the
manuscript is 902, including 34 books, 9 book chapters and 3
technical reports.

Because of the sheer quantity of work published on sustainable investing, it is sometimes imperative to make editing

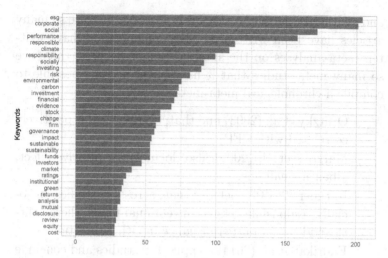

Figure 1.2 **Keywords**. We plot the number of keywords stemming from the references' titles. Only the 35 most frequent words are shown.

choices and cluster many contributions in a short amount of space. Even though it is clearly suboptimal, we occasionally resort to simple lists, instead of literary prose. This is meant to ease the presentation in an exercise that requires concision. Finally, throughout the book, we will resort to jargon and acronyms such as: CSI, or, corporate social irresponsibility which is less frequent than CSR, but sometimes used in studies (see Windsor (2013), Kang et al. (2016) and Price and Sun (2017)), corporate social performance (CSP), corporate environmental performance (CEP), and corporate financial performance (CFP).

1.3 PREVIOUS SURVEYS AND META-STUDIES

The blossoming of ESG investing has generated a thriving literature.[6] There have been several attempts to summarize

---

[6]The present survey focuses on research articles, but many books have been written or edited on this subject. Here are a few in

and synthesize parts of it. In this landscape, because so many articles have been written on SRI, some authors have written meta-analyses on the subject. But as of 2020, there are so many meta-studies that it has almost become possible to compile a chronological meta-meta survey:

- Orlitzky et al. (2003) find that, overall, CSP is positively correlated with CFP.
- Margolis et al. (2009) also document a positive effect, albeit a small one.
- Rathner (2013) concludes that, across 25 studies on socially responsible versus conventional funds, there is little evidence of outperformance in either direction.
- Endrikat et al. (2014) compile 149 studies and conclude that CFP and CEP are generally positively linked (bidirectionally causal).
- Friede et al. (2015) analyze a panel of 2,200 papers and come to the conclusion that 90% of contributions find a non-negative ESG–CFP relation (though this relation may not be significant).
- Revelli and Viviani (2015) find that investing according to CSR purposes is *"neither a weakness nor a strength compared with conventional investments"*.
- Hang et al. (2018) build a large dataset of studies on the CEP-CPF relationship. They perform panel regressions to explain the variations across different variables: countries, GDP, firm size, industries, etc. They find that economic development and legal systems impact the link between ESG performance and firm profitability.
- Mutlu et al. (2018) study the impact of governance on firm performance in China. They indicate that board

chronological order: Domini (2001), Camejo and Aiyer (2002), Little (2008), Krosinsky et al. (2011), Krosinsky and Robins (2012), Gollier (2017), Boubaker et al. (2018), Schoenmaker and Schramade (2018), Sherwood and Pollard (2018), Sachs et al. (2019), Bryant (2020), Edmans (2020), Hill (2020), Bril et al. (2020), Esty and Cort (2020), Taticchi and Demartini (2020), Ziolo (2020) (a collection of thematic chapters), Lukomnik and Hawley (2021), and Silvola (2021).

independence and managerial incentives are financially rewarded.

- Kim (2019) builds a meta-study which reports that many articles find SRI to improve performance, but many papers also point to the opposite; the net effect is that ESG investing is hardly distinguishable from conventional investing, at least from a financial point of view.
- Whelan et al. (2021) compile 1,000 studies published between 2015 and 2000 which assess the link between ESG and financial performance. Based on 245 corporate studies, they report that positive (58%), neutral (13%) and mixed (21%) associations dominate the literature.

In addition to meta-studies, researchers have written thematic reviews. We cite a few of them below in chronological order. The overview of Schueth (2003) provides historical insights on the development of SRI. Beal et al. (2005) review various reasons why individuals may be driven towards ethical investments. Renneboog et al. (2008b) survey various topics, such as regulatory issues, drivers of CSR, and performance of ESG portfolios. An exhaustive list of channels that positively link environmental and economic performance is provided in Ambec and Lanoie (2008). Van Beurden and Gössling (2008) (and Clark and Viehs (2014)) focus essentially on the CSR-CFP relationship. Hoepner (2009) provides a graphical representation (via clusters) of the SRI literature. Chegut et al. (2011) review best practices for ESG investing. Kitzmueller and Shimshack (2012) cover fundamental questions about CSR: why it exists and whether it should exist at all.

Some impacts of SRI are listed in the early review of Wagemans et al. (2013). The survey of Stephenson and Vracheva (2015) focuses on the subtopic of tax avoidance in relation to SRI. Gianfrate et al. (2015) concentrate their analysis on the relationship between sustainability and the cost of capital (see Cantino et al. (2017) for the link with capital structure). On the particular topic of carbon data disclosure, Hahn et al.

(2015) provide a tour of the literature prior to 2015. For sustainable disclosure more generally, we point to Adedeji et al. (2017) and Nwachukwu (2021). The paper by Hjort (2016) exhaustively compiles topics related to climate risks in financial markets (a more up-to-date survey focused on transition risks in Semieniuk et al. (2021)). Hvidkjær (2017) gives a synthetic view of the state of research in ESG investing but concludes that *"the most consistent finding in the current review is that sin stocks exhibit outperformance"*. Brooks and Oikonomou (2018) provide a general purpose review of themes associated with CSR and investments. It is probably the closest in spirit to the present study (along with Piu (2020)).

Slightly more recently, the particular topic of *impact investing* is treated in Agrawal and Hockerts (2019a). Chatzitheodorou et al. (2019) propose a mapping of SRI types and investors motivations. Brakman Reiser and Tucker (2020) survey the landscape of ESG funds: themes, strategies, fees, and voting patterns. Talan and Sharma (2019) propose a classical survey on SRI and identify a few research gaps. Daugaard (2020) provides a recent review that identifies five emerging and trending themes: the human element, climate change, fund flows, fixed income and the rise of non-Western players. Liang and Renneboog (2021) cover topics such as conceptual definitions of CSR, measurement of ESG indicators, and performance of ESG-driven products (equity portfolios and green bonds).

Lately, Camilleri (2020) surveys recent trends of SRI, while Schanzenbach and Sitkoff (2020) focus on legal and conceptual aspects,[7] and Meuer et al. (2020) explore all the definitions of corporate sustainability proposed by the literature. Giglio et al. (2021) and Tokat-Acikel et al. (2021) explore the links between climate change and financial markets (across several asset classes). Likewise, Breitenstein et al. (2021) review practices in environmental hazard mitigation and their impact on

---

[7]We also refer to Fama (2021), in which the topic of CSR is discussed from a contracting perspective.

the financial sector. In the US market, regulatory perspectives linked to the Employee Retirement Income Security Act (ERISA) are provided by Feuer (2020) and Sharfman (2020). A discussion on the comparison with the European equivalent (Institutions for Occupational Retirement Provision - IORP II) is outlined in Daniels, Stevens, and Pratt (Daniels et al.). This matters because individual savings can be nudged toward (or away from) sustainable investments, thereby shifting aggregate demand substantially. Marshall et al. (2021) provide a very clear review of the impact of carbon dioxide emissions on financial markets.

Finally, Kuchler and Stroebel (2020) summarize the field of contributions which link social interactions and financial decision making, and Barroso and Araújo (2021) map and reveal the links (citations) between important papers in the literature. Borghei (2021) surveys the literature on carbon disclosure. Gillan et al. (2021) compile the contributions in corporate finance.

CHAPTER 2

ESG data

One inescapable step in the long journey of SRI is the gathering and processing of ESG-related data. The landscape of ESG data providers is diversified and encompasses both well-established players (e.g., Bloomberg, FTSE, MSCI, Thomson Reuters - Refinitiv, and Standard & Poor's) and more specialized (if not niche) competitors like Sustainalytics, Vigeo EIRIS, GRESB, Carbon4 Finance, RepRisk, Truvalue Labs and Institutional Shareholder Services (ISS), to cite a few. In fact, the field is consolidating: for instance, MSCI has bought RiskMetrics (which owned KLD) and Carbon Delta, while S&P acquired Trucost and Morningstar got hold of Sustainalytics and Moody's of Vigeo Eiris. We refer to Walter (2020) for an overview of the ESG ratings industry and to Escrig-Olmedo et al. (2019) for a review of its recent evolution. Relatedly, the survey of Grewal and Serafeim (2020) gives details on measuring, managing and communicating corporate sustainability performance.

In Figure 2.1, we plot the relative amount of Google queries through time for online searches of the terms *CSR* and *ESG*. It shows that the term CSR has been more popular on average in the past 15 years, but the recent trend is in favor of ESG. One reason for that is that the acronym ESG is progressively becoming mainstream in corporate reporting standards and, consequently, in investors' jargon.

The jargon is typically an aspect of sustainability which

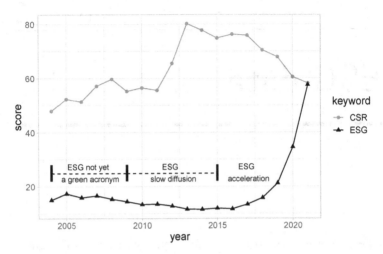

Figure 2.1 **Online queries of keywords through time.** Average annual Google Trends metrics for the terms "*CSR*" and "*ESG*". **Note**: the values for the last year are computed on data up to October 2021.

has thrived. It has become hard to navigate inside the universe of ESG agencies, boards, labels, non-profit organizations, initiatives, principles, regulations, etc., as they encompass a growing number of acronyms. We list a few in Table 2.1.

2.1 OVERVIEW OF ESG ISSUES

2.1.1 Traditional issues

At the company level, ESG data is most often disclosed and analyzed via annual reports. Because there is no regulatory obligation to unveil particular metrics, the amount of information which is available can vary significantly from one firm to another. This variety is such that thousands of fields are reported by ESG data vendors. To simplify a landscape that is hard to fathom for laypeople, it is customary to group indicators by *issues*, and we provide one such categorization in Figure 2.2 (much inspired from Bloomberg). Each E, S and

Acronym	Meaning	Information
BSR	Business for Social Responsibility	A group of experts and firms dedicated to building a sustainable world.
CDP	Carbon Disclosure Project	A not-for-profit charity that helps investors, companies, cities, regions to manage their environmental impacts.
CDSB	Climate Disclosure Standards Board	An international consortium of business and environmental NGOs.
CSRD / NFRD	Corporate Sustainability Reporting Directive / Non-Financial Reporting Directive	A European directive that applies to large firms. The CSRD replaced the NFRD.
GRI	Global Reporting Initiative	An international independent standards organization that helps businesses and governments understand and communicate on their ESG impacts.
IIGCC	Institutional Investors Group on Climate Change	A European membership body for investor collaboration on climate change.
IIRC	International Integrated Reporting Council	A global coalition that advocates sustainable goals for corporate reporting.
SASB	Sustainability Accounting Standards Board	A non-profit organization that seeks to develop sustainability accounting standards.
SDG	Sustainable Development Goals	A blueprint from the United Nations (UN) to achieve a better and more sustainable future for all.
SFDR	Sustainable Finance Disclosure Regulation	A set of EU rules which aim to make the sustainability profile of funds more transparent and comparable.
TFCD	Task Force on Climate-Related Financial Disclosures	A group that develops recommendations for more effective climate-related disclosures.
UNEPFI	United Nations Environment Programme Finance Initiative	A partnership between United Nations Environment Program (UNEP) and the global financial sector to mobilize it toward sustainable development.
UNGC	United Nations Global Compact	A voluntary initiative based on CEO commitments to implement universal sustainability principles.
UNPRI	United Nations Principles for Responsible Investment	An international network of investors working together to implement its six aspirational "*Principles*": 1. To incorporate ESG issues into investment analysis and decision-making processes. 2. To be active owners and incorporate ESG issues into ownership policies and practices. 3. To seek appropriate disclosure on ESG issues by the entities in which they invest. 4. To promote acceptance and implementation of the Principles within the investment industry. 5. To work to enhance their effectiveness in implementing the Principles. 6. To report on their activities and progress towards implementing the Principles.

Table 2.1 List of common ESG and sustainable acronyms.

G pillar is decomposed in a handful of sub-themes which are aggregated to yield a global score (one for each pillar).

Issues Fields (illustrative examples)

E

Climate risk	total greenhouse gas emissions
Emissions & pollution	total SOx, NOx emissions divided by sales
Environmental opportunities	proportion of renewables in energy consumption
Natural capital	total water consumption divided by revenue
Resource efficiency	total waste generation divided by sales

S

Health & safety	incident rate / access to healthcare
Human capital management	equal opportunity policy / whistleblower protection
Product liability	safety fines
Supply chain	human rights policies / audit of supplies

G

Audit	audit committee attendance rate
Corruption & instability	anti-bribery & ethics policy
Diversity	porportion of women executives
Entrenchment	number of board members serving for 10+ years
Independance of board	proportion of independent board directors
Overboarding	proportion of directors in several boards
Remuneration policy	composition of compensation committee
Shareholder rights	poison pill plan

Figure 2.2 **Representative categorization of ESG issues** (and examples of related fields).

Naturally, as the interest in ESG intensifies and firms are incentivized to **disclose** more and more information, the coverage in fields increases. The propensity to disclose ESG information has naturally been shown to be linked to governance (see Michelon and Parbonetti (2012), Ooi et al. (2019), Chouaibi et al. (2021) and Hoang et al. (2021)). Disclosure can be assimilated to an act of good faith and is rewarded (Yoo and Managiy (2021)).

The Carbon Disclosure Project is an initiative that helps investors, companies, cities, states and regions quantify their environmental impact. Each year, they receive information from firms that assess their footprint on the environment. As is depicted in Figure 2.3, the amount of data they have received has increased steadily since the inception of the project in 2003. However, as is also shown, there are major differences in field type. Carbon data is now widely disclosed, but more exotic metrics (e.g., related to forests) are less covered. One reason for that could be that in some sectors, these exotic metrics may not be relevant.

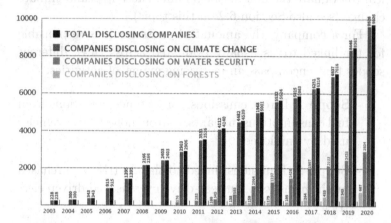

Figure 2.3 **Evolution of coverage**. This graph comes from the Carbon Disclosure Project (CDP - https://www.cdp.net). Each year, the CDP gathers information from firms that voluntarily disclose environmental related data. The above plot shows the increase in data inflow (across several topics) experienced by the CDP.

As is clear in Figure 2.3, climate change metrics are among the fields that are the most reported by companies. However, within these indicators, there are many subcategories. For instance, *climate change* may refer to various elements for a corporation. It can refer to given policies, like having a dedicated team or person in charge of the issue, or adhering to international standards, like those listed in Table 2.1. It can also point to emissions.

There are many emissions that firms can report. For instance, sulfur oxides (SO_x) and nitrogen oxides (NO_x) are highly polluting, but they are not greenhouse gases, which are primarily responsible for global warming. Instead, they may cause acid or toxic rains and are of course hurtful both for the climate and human health (Bilsback et al. (2020)). Greenhouse gas (GHG) emissions, on the other hand, include carbon dioxide (CO_2), methane (CH_4), nitrous oxide (N_2O), and ozone (O_3). The ability to measure these emissions is critical to evaluate the link between human activities and climate change (see also Section 6.2 on this topic).

For a company, the amount of emissions it is responsible for (or linked to) is usually evaluated across three different scopes that encompass an increasing extent of activities:

- **Scope 1**: Direct emissions, for instance, emissions from fuel consumptions in plants, or from non-electric vehicles used by employees.

- **Scope 2**: Emissions linked to the electricity purchased by the firms (e.g., nuclear-driven versus coal-driven electricity).

- **Scope 3**: All other indirect emissions stemming from the value chain, from the supplier (e.g., purchase of raw materials) to the consumer (e.g., use of sold goods). In many cases (i.e., sectors), Scope 3 emissions represent the majority of emissions. Given the amount of stakeholders that are encompassed, Scope 3 emissions are very complicated to assess. There is no dominating paradigm, and Scope 3 values depend strongly on the methodology of the data vendor.

One major issue is that the value chain amounts for a very large portion of emissions, meaning that Scope 3 emissions often dwarf Scope 2 emissions. This is problematic because the bulk of the iceberg is the most important and the hardest to evaluate. We refer to Ducoulombier (2021) for more details on data related challenges of Scope 3 emissions, as well as their integration in portfolio allocation. Cheema-Fox et al. (2021) show that allocations based on Scope 3 emissions differ substantially from those based on more narrow scopes.

2.1.2 Missing data

In the ESG sphere, data providers often scrap information from public sources, such as companies' annual reports. These documents are not standardized with respect to the non-financial fields they are expected to disclose. Therefore, some firms communicate on particular issues, while others do not. This naturally generates heterogeneity in the data that can be accessed.

In order to compute aggregate ESG scores, data vendors rank firms on many criteria, with zero being a low score and 100 the best possible one. The criteria are aggregated into sub-issues, which are compiled into issues, which are, in turn, averaged into the E, S and G pillars. When a given score is not available, the firm is awarded a zero value for the missing field.

Of course, it is possible to resort to imputation methods to replace these missing values. But this always requires to make some distributional assumption on the data. This is likely to impact the outcome of the imputation and may be detrimental if the assumption is incorrect. Sahin et al. (2021) propose another route. Instead of correcting missing points, they add a new pillar, the **Missing** pillar. This pillar computes the proportion of ESG information that is missing for each asset and corrects the proportion to take into account sector specificities (some industries report more than others). It is then easy to integrate this new pillar in portfolio objective functions (see Chapter 5).

2.1.3 Sovereign metrics

Finally, we briefly mention the topic of macro (i.e., sovereign) ESG. At the country level, it is possible to devise indicators that allow ranking nations with respect to sustainable criteria, one of them being the importance of ESG disclosure from a regulatory standpoint (see Singhania and Saini (2021)). Moreover, it seems that in recent periods, disagreement with respect to such measures vanishes. For example, Bouye and

Menville (2021) report that correlation between indicators are often above 80% when comparing alternative rating agencies, e.g., Fitch, Standard and Poor's, and Moody's.

Diaye et al. (2021) reveal positive long-term correlations between sovereign ESG and GDP per capita. Similar results are obtained for national income per capita in Gratcheva et al. (2021). These findings relate to the so-called Environmental Kuznets Curve (EKC) hypothesis, which posits an inverted-U-shaped relationship between pollution and per capita income (see Dinda (2004) for a survey). A stylized version of the curve is depicted in Figure 2.4. The empirical relevance of the curve can be questioned. He and Richard (2010), in their figure 2, show that, in Canada, the relationship between CO_2 emissions and GDP is closer to the pessimistic scenario in Figure 2.4.

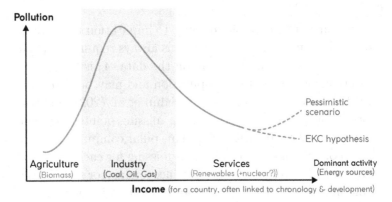

Figure 2.4 **Environmental Kuznets Curve.**

A handful of articles link *sovereign* environmental, social and governance metrics to financial performance. Examples of such issues are:

- **Environmental**: any type of polluting emissions per capita, consumption or production of coal and oil;

- **Social**: life expectancy, income inequality, Human Development Index;

- **Governance**: gender parity in the government, freedom of the press, control of corruption.

For instance, Chang et al. (2020) perform Granger causality tests between stock returns and CO_2 emissions and find that markets Granger-cause emissions, but that the reverse is false. A different facet of this topic is tackled by Morgenstern et al. (2021), who resort to aggregate ESG scores to build macro trend-following strategies. They show that it is financially costless to improve a macro portfolio's exposure to countries that have better ESG ratings.

A different use case of macro data is narrated in Cheema-Fox et al. (2021). The authors show that emerging markets are more vulnerable to physical risks (recorded in the International Disaster Database). Vulnerability is proxied by the Notre Dame Global Adaptation Initiative (ND-GAIN), which evaluates six dimensions: food, water, health, ecosystem services, human habitat, and infrastructure. These results can be exploited to form portfolios of currencies that outperform.

2.2 RATING DISAGREEMENT

The aim of ESG data is to measure and assess the performance of corporations on a large set of issues, which are often categorized according to the environment, social and governance trichotomy. The importance of this data (and its quality) can hardly be overstated, as it drives the decisions of all SR investors around the world. However, with abundance and variety comes heterogeneity: it is not uncommon that rating agencies disagree in their conclusions. This may cause headaches, as a strategy's returns (or any study's conclusions) can very well change when switching from one data provider to another.

This issue is not new (see Griffin and Mahon (1997)), and it is a well-documented inconvenience for portfolio construction, as is for instance shown in Li and Polychronopoulos (2020), Madison and Schiehll (2021) and Schmidt and Zhang (2021): discrepancies in ratings yield diverging performances. Notably, substituting one data vendor for another may shift the investment universe when applying green filters (Billio et al. (2021)). In fact, this lack of consistence may very well deter investors to join the ESG bandwagon (Mooij (2017b)). All is not necessarily lost though: Lopez et al. (2020a) find that ratings seem to agree at least on the worst ranked companies.

An informative starting point is the paper of Widyawati (2020), which compiles a literature review on ESG metrics in which the author shows the complexity of the playing field. It is notably argued that discrepancies in data collection procedures and methodological choices may lead rating agencies to score the same company very differently. Furthermore, Berg et al. (2020) and Abhayawansa and Tyagi (2021) analyze the reasons why different ESG data providers may disagree on a firm's rating. Both articles find that the main drivers of divergence are the scope of categories considered and the measurements methods used. The scope relates to the theoretical definitions that are given to particular fields and topics (e.g., *emissions* or *gender parity*), while the measures pertain to the actual metrics and data sources used to compute these fields. According to Gyönyörová et al. (2021), these divergences depend on countries and industries. In addition to the pure ESG trichotomy, the Sustainable Development Goals (SDGs) are also entering in the objective functions of investors (see De Franco et al. (2021)). Furthermore, Lyon et al. (2018) push even further and contend that firms should also disclose their political activities (party funding notably).

As reported by Dimson et al. (2020a), one other reason for disagreement is the difference in weights that are assigned (by agencies) to the three pillars of ESG. Chatterji et al. (2016) also document the heterogeneity in the conclusions of raters and Kotsantonis and Serafeim (2019) list four challenges that SR investors face when they resort to ESG scores: data inconsistencies, difficulty of benchmarking, risk of imputation and rating divergence (across agencies). As one could expect, disagreement blurs the efficiency of predictive models based on ESG signals. Serafeim and Yoon (2021a) find that scores based on average ESG metrics (across providers) predict future ESG news, but that disagreement (dispersion in metrics) reduces accuracy. Christensen et al. (2021) report that disclosure is also a driver of disagreement and that, surprisingly, higher disclosure leads to *more* disagreement. When rating agencies have a lot of data at their disposal, this may give them more room for interpretation and for data processing.

For theoretical considerations on rating disagreement and its impact on asset prices, we point to Gibson et al. (2021) and Avramov et al. (2021). The latter show that disagreement, when measured by the volatility of the ESG score, can have a strong impact on a firm's alpha because it impacts the effective level of risk aversion. Empirically, Gibson et al. (2021) find that disagreement between agencies is favorable for returns when ratings pertain to environmental measures, but unfavorable when they relate to governance and social issues.

2.3 MEASUREMENT ISSUES AND STABILITY

Beyond quality and trustworthiness, an important property that investors seek in ESG ratings is stability. When investing in best-in-class firms, they both want to reward sustainable policies and to signal that SRI is one of their key priorities. However, if ESG scores fluctuate, it is hard to be confident that the choices that are made at time t remain valid and relevant at time $t+1$. Moreover, if raters reverse their rankings, investors might get confused (Pelizzon et al. (2021)).

For instance, Chatterji et al. (2009) contend that ESG metrics are most of the time reflections of *past* corporate policies but sometimes fail to predict *future* corporate behavior.[1] In fact, it is shown in In et al. (2020) that this depends on the data source and vendor. Utz (2019) confirms that it is very hard to predict corporate scandals from ESG data. Yang

---

[1]There can for instance be mitigating effects across ESG dimensions. Chintrakarn et al. (2020) report that increases in the number of independent directors in boards (which are perceived as good from a governance perspective) decrease the propensity of firms to engage in CSR.

(2021) paints an even darker picture and finds that higher environmental ratings predict increased future unsustainable behavior! In a similar vein, Gidwani (2020) finds that ESG scores are mean-reverting, which complicates the task of ESG-based allocation.

One related issue is the overwriting of scores *a posteriori*. Berg et al. (2021) show that ESG ratings are rewritten through time and that past values are likely to be altered by agencies, so that depending on when the data was downloaded, values may very well change. This is one explanation, among many others, why research may fail to reach a consensus on the impact of ESG on financial, social, and environmental performance.

Finally, we mention that text processing is increasingly used for multiple purposes, from analyzing corporate reports to crafting sentiment indicators. This allows to generate high frequency data, whereas ESG indicators are usually disclosed at annual frequencies. We refer to Cao et al. (2020), Nugent et al. (2020), Raman et al. (2020), Serafeim (2020) and Taleb et al. (2020) for examples of applications.

2.4 GREENWASHING

Greenwashing refers to the practice whereby firms selectively disclose the information that is used to generate ESG ratings.[2] For instance, firms may deceitfully use labels or standards, such as the United Nation's Sustainable Development Goals to signal ethical conduct (see Lashitew (2021)). One way to artificially reduce a firm's carbon emissions is simply to outsource them abroad (Dai et al. (2021)), which impact low scope definitions of emissions, but not third-level scopes

---

[2]Walker and Wan (2012) define greenwashing as the discrepancy between green *walk* (enacting green policies) and green *talk* (communicating about environmental friendly measures). For concise and efficient reviews on greenwashing, we refer to Seele and Gatti (2017), Siano et al. (2017), Falcão et al. (2020) and Yang et al. (2020).

(see Section 2.1). The opposite of this practice is to commit to science-based emission targets (see Freiberg et al. (2021)).

The article by Lyon and Montgomery (2015) offers a survey of such practices and a taxonomy of greenwashing. Typically, one reason why some firms resort to CSR is to signal quality in adverse situations (Gao et al. (2021), Ferrés and Marcet (2021)).[3] The studies by Laufer (2003), Delmas and Burbano (2011), Roulet and Touboul (2015) and Marquis et al. (2016) reveal some drivers of this phenomenon. The second study identifies four types of drivers: market external drivers (e.g., consumers), other external drivers (activists), organizational drivers (incentive culture) and psychological drivers (narrow framing). Roulet and Touboul (2015) (p. 318) conclude that *"paradoxically, beliefs in favor of individual responsibility tend to push firms toward less greenwashing and with a true aspiration to do good, while beliefs in favor of competition lead managers toward egoistic greenwashing strategies based on a gestural commitment only"*. There is documented evidence (Haque and Ntim (2018)) that, often, firms adopt green postures, e.g., by adhering to a sustainable initiative (GRI, UNGC), but fail to improve environmental impact substantively.

Greenwashing is unanimously condemned by all stakeholders in the field of sustainable investing. The reason is clear: by disclosing forged numbers, firms introduce noise in the decision process of investors. This is likely to affect not only the *true* carbon footprint of their portfolios, but also, possibly, the realized financial performance. Delmas and Burbano (2011) categorize firms along two dimensions: their *effective* environmental impact and policies on one hand, and their level of communication on the other hand, see Table 2.2 below.

Counter-intuitively, Marquis et al. (2016) find that corporations that tend to pollute more are particularly likely to eschew selective disclosure, especially if they originate from a

---

[3]Berkovitch et al. (2021) have indeed shown that CSR is a driver of trust for investors.

Communication	True environmental type	
	Brown firms	Green firms
Strong communication	**Greenwashing**	**Vocal green**
No communication	**Silent brown**	**Silent green**

Table 2.2 **Taxonomy of firm communication on sustainability** (from Delmas and Burbano (2011)).

country in which ESG scrutiny is high. They are sometimes those that invest the most in green projects, especially in the energy industries (Cohen et al. (2020)). At the other end of the spectrum, Kim and Lyon (2015) introduce the notion of *brownwashing* (undue modesty, which is not quite equivalent to the silent green firms in Table 2.2). The authors find that a company's growth (endogenous for the firm) and a deregulated environment (exogenous for the firm) are two key variables that influence the decision to indulge in greenwashing or brownwashing.

At the fund level, some managers can be tempted to benefit from ESG labels and expositions without truly improving their ESG footprint (see, e.g., Kim and Yoon (2020)). One notable consequence of such practices is that if past ESG scores are not correlated to future sustainable footprints, then SRI may not be as effective as one could hope (Gibson et al. (2020)). Fortunately, the study of Du (2015) reveals that greenwashing firms end up being penalized once their greenwashing activity is revealed. This phenomenon is negatively associated with financial returns (the study is based on a sample of Chinese firms).

To conclude, we also mention the practice of greenwashing in the financial industry. In Brandon et al. (2021), the authors show that in the US, institutions that publicly commit to responsible investing in fact do not. Some so-called "green" American investors even have worse ESG scores than ESG-independent institutions. Because firms know that they are being watches by analysts on their ESG reporting, they are incentivized to be very careful about the numbers they disclose. In and Schumacher (2021) lay out a very detailed account of how they can alter their performance on carbon emissions.

2.5 GREEN FIRMS

There is obviously no unambiguous definition of *green firms.* Sustainability can be measured in many dimensions and the latter often tell different stories. By default, we are bound to assess *degrees* of greenness.[4] For instance, a major oil producer may pledge to shift its entire research and development (R&D) effort towards renewable energies. This is in fact a reasonable thing to do (see Brown et al. (2021)). By its current situation, the firm is undoubtedly brown, but the direction it is willing to take may attenuate the judgment. Moreover, from an investor's standpoint, funding green patents is somewhat a hedge against risks related to carbon taxes. It seems evident that many firms are now headed in this direction, or at least want the public to think that. In their analysis of the mission statements of European firms, Zumente and Bistrova (2021) find that these statements have evolved, on average, toward more sustainable goals.

Even if green corporations are hard to define, academic studies can again provide some cues. For instance, Diez-Busto et al. (2021) review the literature on the B-Corp movement and detail its main topics and issues: for instance, the factors that lead to certification on one hand, and the social efficiency of B-Corporations on the other.

The motivations that drive companies towards sustainability are well documented. In their seminal article on CSR, McWilliams and Siegel (2001) mention various factors that are likely to drive a firm's appetite for sustainability. The abstract mentions the following attributes: size, level of diversification, research and development, advertising, government

---

[4]Which is exactly what data providers do by selling scores and ratings on a quasi-continuous scale, often from 0 (worst) to 100 (best).

sales, consumer income, labor market conditions, and stage in the industry life cycle.

Nevertheless, governance seems to be the focal angle in academia and the composition of boards is a recurring subject.[5] Hoang et al. (2021) find that some board characteristics matter for environmental performance (CEO-chairman duality, gender diversity, board member age), while others don't (board size). Velte (2016) and Lee and Kim (2021) report that having more women on boards implies better ESG performance, or mitigates the risk of brown policies. Indeed, as Haque (2017) conclude, board independence and board gender diversity are positively linked with carbon reduction initiatives. Moreover, climate disclosure increases with female representation (see Ooi et al. (2019)). From a valuation standpoint, this can be beneficial to firms, as Kang et al. (2021) report that so-called activist directors (appointed through shareholder activism events) are linked with large firm value increases.

However, the study of Moussa et al. (2020) attenuates these findings by revealing that board characteristics only play a role (in the reduction of pollution) *if the firm has a carbon strategy*. Relatedly, Bento and Gianfrate (2020) show that gender representation among directors and independence of board members is linked to internal carbon pricing (which penalizes intensive carbon activities), but so are other exogenous factors, such as revenues, profitability, and GDP per capita of the firm's country. The study Bu et al. (2021) introduces lesser known concepts. First, it focuses on the role of talented inside directors (TIDs), who are inside directors with outside directorship. Second, the dependent variable is the residual CSR that cannot be attributed other factors, such as advertisement, R&D, cash flows, etc. Surprisingly, the authors call this residual the *excess* CSR and show that the presence of TIDs negatively relates to excess CSR. The authors conclude that TIDs are beneficial for firms because they prevent CEOs

---

[5]Another factor: CSR can be driven by the local social capital surrounding the firm (see Jha and Cox (2015)).

from engaging in too much CSR (thereby implying that there is such thing as too much CSR).

Whether so-called green firms are actually green (or remain green) is another topic. This relates to the stability of ratings mentioned in Section 2.3. When a firm is rated as highly environmentally friendly, we should anticipate the rating to stay high, and, relatedly, we should expect green policies from the top management. However, in their staggering study, Elmalt et al. (2021) reveal that firms with higher ESG scores are only *weakly* linked to lower carbon emissions. A more positive note comes from Saeed et al. (2021) who find that, in the energy sector, firms with a CSR committee pollute less.

2.6 AD-HOC SOLUTIONS

First of all, in order to circumvent the obstacle of divergent ratings, some investors have decided to build their own in-house metrics (see Serafeim and Grewal (2017)). A second way to mitigate divergence is aggregation: Jacobsen et al. (2019) advocate ratings combination, which reduces dispersion and measurement errors. A similar attempt in this direction is the work of Gibson and Krueger (2018), which introduces an aggregate metric (at the portfolio level) of sustainability footprint. This is also intended to help investors measure the ESG impact of their investments.

Another issue is missing data, which often occurs when smaller firms do not disclose any ESG-related information. In this case, Henriksson et al. (2019) argue that it is possible to construct an ESG long-short factor and evaluate which individual stocks load heavily on it. This allows to build a proxy for ESG exposure. Görgen et al. (2020) and Roncalli et al. (2020, 2021) also follow this route and construct a brown minus green (BMG) factor so as to assess a firm's exposure via the beta coefficient with respect to this factor. It is then possible to score firms even when they do not disclose data related to their sustainability policies. Whether such metrics are faithful reflections of ESG standards remains an open question.

In a different direction, Gaganis et al. (2021) propose that banks use their unique position to craft in-house indicators. They argue that banks should evaluate corporate borrowers along ESG lines and their paper proposes a methodology to do so.

We end this subsection with a short list of further contributions. To mitigate some of the aforementioned issues, Fiaschi et al. (2020) propose a quantitative (quantile regression based) solution to improve the measurement of corporate wrongdoing. Another stream of the literature examines the drivers of the propensity to disclose ESG data. For example, Rezaee and Tuo (2019) argue that ESG reporting is linked to the quality of earnings (while Jia and Li (2021) link corporate sustainability to the *persistence* in earnings). Other useful insights and pieces of advice on the handling of ESG data and its integration in the investment process can be found in Hallerbach et al. (2004), Kotsantonis et al. (2016), Bender et al. (2017), Bender et al. (2018) and Cornell and Damodaran (2020). Finally, we end this subsection by mentioning the work of Mahmoud and Meyer (2020). The authors decompose the ESG ratings along three orthogonal axes: uncertainty, investor sentiment, and idiosyncratic sustainability. In their analysis of the COVID-19 crash, they show that the immunity of sustainable stocks is linked to the uncertainty factor more significantly than to the core sustainability axis.

2.7 NEED FOR TRANSPARENT AND UNIFORM REPORTING

The most efficient solution is probably when governments impose ESG reporting obligations, even though, in some cases, firms may find advantages to disclose their indicators and metrics in the absence regulatory constraints (Ioannou and Serafeim (2019)). In fact, some companies write corporate reports in such ways that they are favorably analyzed by automatic algorithms (Cao et al. (2020)), for instance by optimizing textual sentiment. In any case, ESG reporting will become

paramount because investors will ask for more transparency
(Rissman and Kearney (2019)). Because of the complex nature of all ESG dimensions, Kaplan and Ramanna (2021) suggest to start by focusing on the most consensual fields, such as
emissions and indentured labor in supply chains. Krueger et al.
(2021) for instance document the benefits of mandatory disclosure, and as expected, it increases the availability and quality
of ESG reporting. Recent advances in non-financial reporting
are reviewed in Pronobis and Venuti (2021) and Harper Ho
(2021), the latter being focused on the US.

This is all the more critical that easily found information
is never as valuable as sophisticated and esoteric data (Bose
(2020)). The simplest way to achieve this goal is to impose
ESG reporting standards (Lopez et al. (2020b)). Currently,
large firms, because they can devote more resources, have an
edge with respect to ESG disclosure (Artiach et al. (2010),
Drempetic et al. (2020)). Moreover, according to Chen et al.
(2021), *"the amount of ESG data that is available for a given
company is positively correlated with the commercial ESG rating of that company, and also the weighted average cost of
capital for that company"*.

The drivers of voluntary disclosure remain largely un-
known, but the studies of Ooi et al. (2019), Yu and Van Luu
(2021) and Chouaibi et al. (2021) seem to indicate that some
governance characteristics play a role (e.g. size, independence,
and diversity). Some initiatives have already blossomed, and
they promote transparency and exhaustiveness in reporting,
like the Task Force on Climate-related Financial Disclosures
(TFCD),[6] and efforts in this direction need to intensify, as is
advocated in the roadmap of the Board (2021).

However, there exists a risk of *rating addiction* (Cash
(2018)): it is imperative not to fall in the trap of rating agencies and make sure that ratings are fair, rational and independent from corporate funding and conflicts of interest (Hoepner and Yu (2017b)). Tang et al. (2021) for instance find that

---

[6]See https://www.fsb-tcfd.org and https://www.tcfdhub.org.

raters tend to favor firms which share the same parent company. Sadly, transparency may not be in the interest of these agencies because their method is a large part of their intellectual property (Stubbs and Rogers (2013)).

CHAPTER 3

Investors and SRI

This chapter is dedicated to investors: what their preferences, beliefs and practices may be, and how these have changed recently. On the narrower topic of institutional investors, we recommend the well-documented surveys of Matos (2020) and Crifo et al. (2019) (for a focus on France in the latter).

To illustrate the growing appetite of investors for ESG assets, we plot in Figure 3.1 the value through time of assets considered to belong to ESG criteria in the US. The recent steep increase mirrors and echoes that of Figure 2.1. In order to quantify the aggregate appetite for sustainable firms and funds, Briere and Ramelli (2021a) develop a green sentiment index by calculating monthly abnormal flows into environmentally friendly exchange traded funds (ETFs).

3.1 INVESTOR PREFERENCES AND BELIEFS

A central question in SRI is: why do people allocate to ESG-driven assets?[1] The simplest answer is that asset owners care not only about pure profitability, but also about their ESG

[1]There are different levels of granularity for assets: at the aggregate level, people can choose to invest in sustainable funds versus conventional funds. At a more micro-level, it is also possible to discriminate firms according to their ESG practices and build portfolios accordingly.

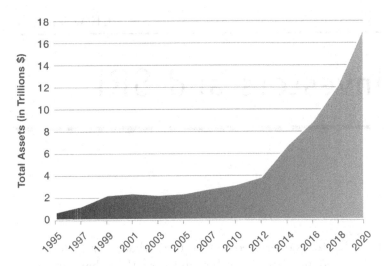

Figure 3.1 **Evolution of ESG assets under management**. (US SIF Foundation (Forum for Sustainable and Responsible Investment). **Via**: Report on US Sustainable and Impact Investing Trends 2020.)

footprint.[2] In fact, according to Amel-Zadeh (2021) experts and investors seem to agree that climate change is financially

---

[2] A more complete answer is that there are two separate reasons:

Socially responsible investors are driven by $\begin{cases} \text{social/environmental good} \\ \text{pecuniary incentives} \end{cases}$

The first reason is straightforward and corresponds to the case where the investor has ethical concerns and genuinely cares about the future state of the world. The second motivation can have several roots. For many reasons, sustainable firms can be perceived as less risky and/or more profitable in the long run (see Fernando et al. (2017)). It is often assumed that firms with high ESG scores (especially high E scores) are hedges against climate related risks (see Chapter 6). Also, they may benefit from more favorable tax dispositions, compared to brown companies. The (possibly unbalanced) mixture of these two inspirations is the source of ESG investing. For investors, this can however be a source of headaches. Dual (and possibly contradictory) incentives in the ESG versus performance dimensions complicate the task of attracting flows (see Gantchev et al. (2020)).

material, but most firms do not believe they are exposed to weather-related risks. This creates an asymmetry in the assessment of exposures which is mostly explained by the lack of data and methods that evaluate these risks.

While this section is dedicated to *green* investors, it is also noteworthy to fathom the characteristics of *brown* investors. In their study on tobacco stock ownership, Blitz and Swinkels (2021b) show that investors in tobacco stocks are on average more often anonymous, compared to investors in peer stocks.

In their experiment involving Swedish investors, Lagerkvist et al. (2020) find that fund preferences are more driven by sustainability criteria, than other characteristics (like fees, fund type, risk level, or geographical focus). According to another natural experiment carried out in Hartzmark and Sussman (2019), investors do value sustainability and derive utility both from financial returns *and* social returns. Finally, in a third natural experiment, Chew and Li (2021) reveal that individuals exhibit both *sin stock aversion* and *virtue stock affinity*. However, there is an asymmetry: avoiding "bad" investments matters more than investing in good (i.e., green) ones.

In addition, Bollen (2007) and Barber et al. (2021) also find that investors derive non-pecuniary utility from investing in funds that are not only profitability-driven. More precisely, Mahmoud (2020) argues that SRI can be fostered by mechanisms of warm glow or self-image, or simply by personal happiness. This is a two-edged sword because, as Heeb et al. (2021) show, the so-called *impact* investors can be tempted to optimize this warm glow instead of their true impact *per se*. Humphrey and Li (2021) reveal that mutual funds reduce the carbon impact via two channels: either because they have signed the Principles for Responsible Investment (thereby bending their policies toward greener firms), or simply because they have environmentally-aware stakeholders. The recent trend does indeed confirm that mutual funds in the US have shrunk their holdings in black (carbon-intensive) asset by almost a factor 2 between 2012 and 2018 (Muñoz (2021a)). Relatedly, Alda (2021) reports that conventional firms are more and more tilting their holdings toward ESG firms so that they are slowly converging to their green counterparts.

One way to measure the aggregate appetite for SRI is to use internet search volume on this topic. Chen et al. (2019) find that firms that are positively exposed to this social sentiment proxy attract higher institutional demand and earn positive abnormal returns. The COVID-19 crisis has cemented and accelerated the demand for sustainable assets and funds. For Pástor and Vorsatz (2020), sustainability has become "*a necessity rather than a luxury good.*" Other sources of information are company filings and Iliev et al. (2021) show that while investors do monitor firms' governance, there are discrepancies because much of the focus is on large firms (this is confirmed in Azar et al. (2021)). When analyzing ESG news and fund flows, Chen et al. (2020) also find that active managers react to ESG information because they seek to cater to some of their clients' demands.

Because ESG-driven investors are less interested by financial performance, Benson and Humphrey (2008) and Renneboog et al. (2011) find that SRI fund flows are less sensitive to past returns, compared to flows coming in and out of conventional funds.[3] This may explain why, more and more, mutual funds compete to obtain eco-friendly certifications (Ceccarelli et al. (2020)), if only because they are vectors of differentiation, which some funds may seek (Rankin (2020)). However, these green labels are by no means a guarantee of ex-post ESG performance by the fund portfolios (Kim and Yoon (2020)). One reason may be that even when CEOs are incentivized toward sustainable goals, reaching these goals may take time (Derchi et al. (2021)).

Another reason why institutions turn to SRI comes from their customers or other stakeholders (Majoch et al. (2017)). Amel-Zadeh and Serafeim (2018) show that beyond performance, investors resort to ESG because of client demand and

---

[3]ESG-driven institutional investors are also less reactive to quantitative mispricing signals (e.g., standardized unexpected earnings), see Cao et al. (2020).

product strategy (see also Jagannathan et al. (2018)). Finally, environmental awareness and financial performance may be linked, because, as Cormier and Magnan (1997) explain, investors are afraid that polluting companies will be sanctioned, thereby incurring a financial cost and lowering returns. Indeed, because of climate change risks (see Chapter 6), investors anticipate more regulatory constraints in favor of the environment, which will negatively impact the most polluting firms (Ramelli et al. (2020)). One such example is the carbon tax which can be quite costly for some industries (Bertolotti and Kent (2019), Carattini and Sen (2019)).

Since green investors ask less return for firms that make efforts toward sustainability, SRI is found to be negatively related to cost of equity (Sharfman and Fernando (2008), El Ghoul et al. (2011) - see also Section 4.7). Similarly, Chava (2014) find that investors demand significantly higher expected returns on stocks excluded by environmental screens. Likewise, Stotz (2021) finds that discount rates of high-ESG stocks have fallen relative to low-ESG stocks, so that high ESG stocks have higher realized returns, but lower *expected* returns. One reason for that, as Davis and Lescott (2019) argue, is that all other things equal, a firm with lower ESG scores has riskier cash flows, which pushes its cost of capital up. Indeed, some stockholders view ESG-driven firms as more resilient (Roselle (2016)).

The inclination toward sustainability can further be elucidated from social perspectives (Hoepner et al. (2019)). According to Riedl and Smeets (2017), investors hold socially responsible mutual funds because of social preferences and social signalling. Typically, Democrats hold sin stocks in smaller proportions, compared to Republicans (Hong and Kostovetsky (2012)), and Republican shareholders are related to firms with lower environmental performance (Kim et al. (2020)).[4]

---

[4]Relatedly, Bolton et al. (2020) are able to rank institutional investors on a left-right scale by analyzing a broad dataset of voting records. Public pension funds are found to be on the left of the axis and are more likely to support ESG policies. In Muñoz (2021b), it is found that

Investors who view CSR as beneficial to society naturally tend to favor ESG firms (Arnold et al. (2020)). Local pollution is also found to be a driver of investors' preference: Huynh et al. (2021) find that managers who live in polluted areas tend to underweight firms with high emissions in their portfolios.

Religiosity also seems to be positively correlated with the propensity to invest in sustainable companies (Bae et al. (2019)), or to the propensity toward ESG disclosure (see the study of Terzani and Turzo (2021) which operates at the country level and Muñoz (2021b) for a discussion on sin sectors). Some mutual funds are even dedicated to religion-based SRI (see Stultz (2020)). Relatedly, Glac (2012) finds that the integration of investment frames (i.e., when social beliefs are important in the investment domain) are positively correlated with propensity to indulge in SRI. In a similar vein, norm-constrained institutions such as pension plans are less likely to hold sin stocks, compared to mutual or hedge funds (Hong and Kacperczyk (2009)). Overall, peer effects, such as those uncovered in Karakostas et al. (2021) are probably at play in the propensity to invest in green assets.

What are some other characteristics of ESG investors? First of all, SR investors are more loyal: Bollen (2007) finds that investor cash flow volatility is lower for SRI funds than for conventional mutual funds. There are also mimicking effets: Hellström et al. (2020) find that parents' and children's SRI behavior are correlated. Humphrey et al. (2021) document asymmetries in preferences: individual investors are more inclined to avoid negative ESG externalities than they are to embrace the positive externalities. While mutual funds have been active drivers of SRI, ESG-based private equity is now also thriving (Indahl and Jacobsen (2019), Zaccone and Pedrini (2020), Alfonso-Ercan (2020) and Long and Johnstone (2021))[5]. This is also true for venture capital and we refer to

---

mutual funds managed in Democrat-leaning states are less exposed to sin sectors.

[5]In addition to academic references, practitioner surveys also document this trend. We refer, e.g., to PriceWaterhouseCoopers (*Private*

Block et al. (2020) and Barber et al. (2021) for studies on the preferences of impact investors targeting young companies. However, appetite for SRI is not restricted to institutional investors. Barreda-Tarrazona et al. (2011), Kim et al. (2018), Rossi et al. (2019) and Bauer et al. (2021) document a recent increase of interest from households and retail investors. Individuals are indeed sensitive to non-financial data when crafting investment decisions (Mervelskemper (2018)). However, they do not process the ESG information like institutional investors (Pelizzon et al. (2021)).

Shareholders are increasingly worried about the impact of climate change (see also Chapter 6). According to Krueger et al. (2020), investors believe climate change will have repercussions on equity valuations which should be hedged via risk management and engagement. This is true irrespective of the political environment. Ramelli et al. (2021) report that even under the (rather polluter friendly) Trump administration, investors continued to reward sustainable firms.[6] However, as is noted in Chapter 2, the lack of homogeneity and transparency in ESG reporting makes it hard to accurately assess firms' social and carbon footprints. Ilhan et al. (2021) report that many investors *"think climate risk reporting to be as important as traditional financial reporting and that it should be mandatory and more standardized."*

We end this section by underlining the importance of nudging in SRI. Briere and Ramelli (2021b) document the significant impact of offering a sustainable investment option in employee saving plans (in France). Naturally, providing the option of choosing a sustainable fund increases the participation of retail investors to the green finance movement.

Equity Responsible Investment Survey 2019) and the Intertrust Group (*Global Private Equity Outlook 2020*).

[6]A historical note: early into the Biden administration, in February 2021, the Interagency Working Group on Social Cost of Greenhouse Gases released a technical document entitled, *Social Cost of Carbon, Methane, and Nitrous Oxide.*

3.2 IMPACT INVESTING

Another focal question in SRI is the effect that it has on corporations' policies and, in turn, on the social sphere.[7] There is some disagreement in the literature on whether investors are able to influence corporate policies efficiently. For instance, Coffey and Fryxell (1991) report positive impacts with respect to gender parity, but no influence on charity giving. Using institutional holdings combined to a sophisticated demand estimation technique, Noh and Oh (2021) find that investors are able to put pressure on firms – willingly or not. This institutional pressure for *greenness* is positively linked to firms' future environmental performance. Relatedly, Dikolli et al. (2021), in their study on close to 4,000 shareholder proposals, conclude that ESG mutual funds are more likely than non-ESG mutual funds to support environmental and social (E&S) shareholder proposals.

In contrast, David et al. (2007) conclude that shareholder activism may in fact be counter-productive and lead to opposite effects. In Cohen et al. (2020), the authors show, that paradoxically, brown firms are much more invested in green R&D, especially in the oil and gas industries. Thus, screening them out of ESG portfolios is not supportive of sustainable research and patenting. In addition, few so-called "impact funds" tie the compensation of their managers to the (blurry) notion of impact. The incentives are thus not aligned with the purpose of the funds, and the measurement of their impact is often foggy.[8] What's worse, de Groot et al. (2021)

---

[7]The line between SRI and impact investing is blurry. We refer to Roundy et al. (2017), Agrawal and Hockerts (2019a,b), Block et al. (2020), Cojoianu et al. (2021) and Bengo et al. (2021) for discussions on this topic and on the profiles of the so-called *impact investors*. For a historical perspective, we recommend the chapter by Martin (2020).

[8]Interestingly, CSR can be framed as an **agency problem** (Barrios et al. (2014), Martínez-Ferrero et al. (2016), Li (2018), Liang and Renneboog (2018), Hussaini et al. (2021)), though this perspective is partly rebuked in Ferrell et al. (2016). Relatedly, see Haque and Ntim (2020) for a study that links executive remuneration to carbon performance and firm value.

show that portfolio managers predominantly vote against social and environmental proposals. Even those who adhere to the Principles for Responsible Investment are not inclined to support green policies. One possible reason is that asset managers are not accustomed to analyze non-financial data (Arjaliès and Bansal (2018)), or may be confused by it (Pelizzon et al. (2021)). This is of course less and less plausible, as investors are increasingly concerned with ESG issues. Serafeim and Yoon (2021b) find that investors react to ESG news, especially if the latter relate to social capital and are susceptible to impact a firm's fundamentals. Overall, this resonates with a warning from the theoretical paper by Yan et al. (2019) which warns that "*the relationship between the dominant financial logic and the social logic of SRI shifts from complementary to competing as the financial logic becomes more prevalent in society and its profit-maximizing end becomes taken for granted.*"

The middle ground (see Simerly (1995) and Heath et al. (2021) for instance) establishes that investors do not significantly impact corporate policies, or that the impact is contingent on investor type (hedge fund, bank, mutual fund; see Johnson and Greening (1999)). In Blitz et al. (2021) it is found that even though sustainable investing increases green firms' access to capital, it does not (yet) limit funding opportunities for non-sustainable companies. Also, success of shareholder activism may be conditional on some favorable factors. Barko et al. (2021) document that activism is "*more likely to succeed when targets have a good ex ante ESG track record, lower ownership concentration and growth.*"

The recent trend of research seems to nonetheless establish *some* effect. It may be direct: large shareholders command[9] that the firm enforces sustainable choices (lower carbon emissions, development of environmentally friendly

---

[9]This can be done, e.g., via incentives, see Ikram et al. (2019), Li et al. (2020), Li and Thibodeau (2019) and Dunbar et al. (2020), or via direct votes (Curtis et al. (2021)). For a theoretical perspective, see Morgan and Tumlinson (2019).

products, board gender parity (see Ghosh et al. (2016)).[10] Other repercussions are indirect: witnessing the money flows toward SR firms, executives may be inclined to orient their policies to match ESG goals so as to capture some of these flows, or to improve the media coverage of their companies (Cahan et al. (2015)).[11] In fact, it has been shown (Kang et al. (2021)) that signalling sustainability is one way to increase capitalization: when firms are included in green indices, the demand stemming from investors increases, which pushes prices up. Reversely, controversies reduce the odds of being included in green indices (Arribas et al. (2021)). In addition, negative media coverage pertaining to sustainability increases the chances of CEO ouster (Colak et al. (2020), Burke (2021)), which incentivizes the top management to act responsibly. We refer to Gillan and Starks (2000), Gillan and Starks (2007), Goodwin (2016), Zeng and Strobl (2016) and Gomtsian (2020) for studies on shareholder activism and the role of institutional investors. Bebchuk and Hirst (2019) reviews the impact of index funds on corporate governance.

Kölbel et al. (2020) review the mechanisms of SRI's impact. They list three channels (shareholder engagement, capital allocation, and indirect impacts) but notice that only the first one is well supported by the literature. One such contribution is the work of Dyck et al. (2019) which shows that *"institutional investors transplant their social norms regarding E&S issues throughout the world."* The authors find that

---

[10]Board composition is expected to have an effect on policies (Moussa et al. (2020)). However, this impact is not straightforward, as Haque (2017) reports that board composition can shift the willingness to curb carbon emissions, but not necessarily on the greenhouse gas reductions per se.

[11]For instance, Francoeur et al. (2021) document that powerful CEOs of polluting companies are able to reduce the environmental footprint of their firms. In addition, Lee and Kim (2021) find that overconfident CEOs invest more in ESG policies in South Korea. We refer to Huang et al. (2021) for a study on the impact of star CEOs on ESG performance in China. With regard to CEOs in general, Wernicke et al. (2021) find that they can explain 30% of the total variance in CSR.

these investors care more about ESG after-shocks that reveal how valuable sustainability is (e.g., market crashes or climate-related disasters). Li et al. (2020) find two practices that mutual funds use to impact firms' ESG policies: voting and CEO compensation (i.e., ESG incentives). Their findings are more marked when board governance is strong and when the funds hold significant shares of the companies. Governance is often key. Saeed et al. (2021) find that CSR governance is an efficient driver of reductions of carbon emissions.

A related stream of the literature pertains to ownership. Shive and Forster (2020) document that public firms owned by mutual funds have lower carbon emissions. Kordsachia et al. (2021) find that sustainable institutional ownership (measured by the signatory status to the UN Principles for Responsible Investment) is positively associated with a firm's environmental performance. In addition, Cox et al. (2004), Neubaum and Zahra (2006) and Kim et al. (2019) find that the relationship between institutional ownership and ESG outcomes is more salient when the investment horizons increases (and, as McCahery et al. (2016) show, investors intervene more when they have long-term perspectives). Similar results are derived in Boubaker et al. (2017), Starks et al. (2020), Kim et al. (2018) and Gloßner (2019). In contrast, short-term investors tend to prefer polluting firms (Tirodkar (2020)), and they have the most influence on firm values because they discipline managers through credible threats of exit (Döring et al. (2021)).

The effectiveness of environmental activist investing is nonetheless still debated. Naaraayanan et al. (2021) report that targeted firms end up reducing their toxic releases, greenhouse gas emissions, and cancer-causing pollution. Some researchers however are less optimistic about the ability of SRI to impact companies' policies. Blitz and Swinkels (2020) contend that excluding unsustainable stocks from portfolios does not contribute to making the world a better place because the corresponding firms are then owned by investors who do not care about ESG issues (and thus do not bend corporate policy towards sustainable objectives). One example is detailed in Fu et al. (2020), where it is shown how ESG-driven share-holders can impact the gaming industry. Also, ESG concerns must be shared both by shareholders and top management. When the latter is not supportive of ESG policies that the former push, firms may be negatively impacted by negative incidents (He et al. (2020)), the latter being taken into account by analysts when they update their earnings forecast (Derrien et al. (2021)). Finally, DesJardine et al. (2021) find that activist hedge funds are a threat to sustainability because the targeted firms end up curtailing SR policies. In fact, the complexity of ownership structure may sometimes blur our ability to understand which investors own shares and voting risk in brown firms (Mizuno et al. (2021)).

Nevertheless, other channels may also be at work. Indeed, Dai et al. (2021) find that corporate customers can impact the CSR policy of their suppliers. Large buyers, because they are strong drivers of business, have some power to curb their suppliers' behaviors. These buyers do not want to take the risk of bad publicity stemming from an association to a faulty partner. The reverse effect, whereby suppliers affect the risk of their clients is documented in Chen et al. (2021). Though not yet documented in the literature, these phenomena are probably amplified by the importance of social networks. One example of a social movement that impacted SRI in France is documented in Arjaliès (2010). Retail customers are also sensitive to ESG-driven companies (Radhouane et al. (2018)).

We end this section with a single number. Berk and van Binsbergen (2021) prove an approximate formula for the impact of impact investors on the cost of capital of firms. The difference in cost of capital with and without socially responsible investors is found to be equal to $\mu f(1 - \rho^2)W_e/W_{-e}$, where μ is the market risk premium, f is the proportion of the economy that is targeted by the investors, ρ is the correlation between green and brown firms, and W_e and W_{-e} is the wealth dedicated to green investments versus the rest of the wealth. The authors, with historical figures, find that this number equals 0.35 basis points because W_e is only about 2%.

This number is therfore too small to impact real investment decisions.

3.3 INVESTOR PRACTICES

Once the decision is taken to invest in green firms, one question remains: how? While there are many ways to proceed qualitatively and quantitatively (see Mooij (2017a) as well as Chapter 5), we briefly mention some simple routines below. In their survey of investor practices, Van Duuren et al. (2016) show that practitioners use ESG data for red flagging and risk management. In another such survey, Eccles et al. (2017) report that practitioners use ESG data essentially for screening purposes or best-in-class selection. This is confirmed in Cox et al. (2004) as well as in Hoepner and Schopohl (2018) for two very large institutional investors. In addition, they conclude that one of the main arguments for ESG investing is that it helps *foster a long-term investment mindset.*

According to Dorfleitner et al. (2021), highly remunerated asset managers have an asymmetric posture with respect to sustainability. While they react rapidly to controversies (and avoid firms hit by scandals), they are much less pro-active with respect to corporate green initiatives. Nevertheless, as is outlined in Rhodes (2010), screening can be hard to perform efficiently.

We end this section with a pot-pourri of practices described in the literature. Nofsinger et al. (2019) report that investors underweight stocks with negative E&S indicators, likely because they associate negative E&S scores with downside risk. Moss et al. (2020) find that retail investors do not respond to ESG disclosures, in contrast to institutional investors, who seem to do. Berry and Junkus (2013) show that investors seem to prefer best-in-class to negative screening, thereby rewarding green firms but not penalizing unsustainable ones. Matallín-Sáez et al. (2021) study investor flows in and out of mutual funds. They find a striking pattern. For conventional funds, outflows correlate negatively with past performance (e.g., losses are penalized, which increases outflows, gains are rewarded, which reduces them). For socially responsible funds, it is the opposite: investors hold on to green funds that have underperformed.

Kim and Yoon (2020) introduce the notion of fake ESG investing, that is, when some fund tries to attract capital via ESG labelling but fails to improve its ESG footprint afterward.

Atz et al. (2019) propose a five-step methodology to assess what they call *ROSI* (return on sustainability investments). The steps start by identifying strategies and actions and end by monetizing and evaluating their consequences. Finally, according to Adriaan Boermans and Galema (2020), investors are subject to home bias. When they originate from countries with many carbon-intensive firms, their portfolios are logically more exposed to such types of firms.

CHAPTER 4

ESG investing and financial performance

One of the central questions in ESG investing pertains to its opportunity cost: is sustainable investing penalizing from a pure pecuniary standpoint? This chapter addresses this complex and highly debated question. We start by a simple model that illustrates some typical mechanisms that can enlighten the situations when green firms can outperform brown ones. Then, we split the contributions of the impact of ethical screens on financial performance into four categories: the ones that document a positive relationship, the ones that find a negative relationship, the ones that report no particular relationship, and finally the ones that conclude that the relationship is contingent on some external factors. We mostly refer to financial performance in the sense of returns, though sometimes it means valuation. Nevertheless, SRI impacts other aspects of corporations' finances, with risk being a major concern. We list some of these facets in the last two subsections below.

4.1 TOY MODEL

The mathematics-averse reader is advised to skip this subsection.

4.1.1 Theory: assets, agents, equilibrium

A full section of the survey is dedicated to theoretical models (the online Chapter), so the framework we present here is
very stylized.[1] We consider two agents and two risky assets
(for simplicity). The two assets are indexed by g for *green*
and b for *brown*, while the two agents are indexed with up-
percase letters G (sustainability sensitive agent) and B (less
ESG-driven, or even ESG insensitive agent). We will often use
X to denote one of the agents (G or B) and y one of the assets
(g or b).

Assets. Prices are denoted with p_g and p_b. Both assets are
expected to yield a cash-flow (or payoff): z_g and z_b, respectively, and we assume that they have a correlation equal to
ρ. Agents agree on the variances of these payoffs, σ_g^2 and σ_b^2.
However, they are allowed to disagree on their means: agent
G estimates they are m_{Gg} and m_{Gb} while agent B believes
they are m_{Bg} and m_{Bb}. The assets are also characterized by
a non-random (observable[2]) ESG score: e_g and e_b, such that,
naturally, $e_g > e_b$. Both assets have fixed supplies s_g and s_b,
which are supplied by so-called *noise traders* who trade for
reasons other than payoffs. One risk-free asset is also available in unlimited supply, with unit price and certain payoff
equal to $r > 0$.

Agents. Agents have initial wealth W_G and W_B, for the *Green*
(sustainability-driven) and *Brown* agent, respectively. Each
agent $X \in \{G, B\}$ buys a quantity q_{Xy} of shares of asset
$y \in \{g, b\}$. The terminal wealth of agents satisfies

$$W_X^* = \underbrace{r(W_X - q_{Xg}p_g - q_{Xb}p_b)}_{\text{payoff from riskless asset}} + \underbrace{q_{Xg}z_g + q_{Xb}z_b}_{\text{payoff from risky assets}}, \ X \in \{G, B\}.$$

[1]It is inspired from the model of Kacperczyk et al. (2019), though
the utility function and the market clearing mechanism are different.

[2]Again, this is a strong assumption because in practice, ratings disagree, see Section 2.2. The article Avramov et al. (2021) proposes a
theoretical model with disagreement.

We re-write this expression in the form of the gross return:

$$\frac{W_X^*}{W_X} = r(1 - w_{Xg}p_g - w_{Xb}p_b) + w_{Xg}z_g + w_{Xb}z_b, \quad X \in \{G, B\},$$

where $w_{Xy} = q_{Xy}/W_X$ is the number of shares of agent X in asset y, divided by the total initial wealth of the agent.[3] Also, the ESG score of agent X's holdings is evaluated as

$$E_X = w_{Xg}e_g + w_{Xb}e_b.$$

The agents have a quadratic utility on gross return on wealth, plus an ESG component:

$$U_X = \mathbb{E}\left[\frac{W_X^*}{W_X}\right] - \frac{\gamma}{2}\mathbb{V}\left[\frac{W_X^*}{W_X}\right] + \delta_X E_X,$$

where $\delta_G > \delta_B$ means that agent G cares more about ESG score than agent B. The risk aversion coefficient $\gamma > 0$ is common to both agents. If we use bold vector notations for prices $\boldsymbol{p} = [p_g \ p_b]$ and relative portfolio holdings $\boldsymbol{w}_X = [q_{Xg} \ q_{Xb}]$, then the utility function reads

$$U_X = r - r\boldsymbol{w}_X'\boldsymbol{p} + \boldsymbol{w}_X'\boldsymbol{m}_X - \frac{\gamma}{2}\boldsymbol{w}_X'\boldsymbol{\Sigma}\boldsymbol{w}_X + \delta_X\boldsymbol{w}_X'\boldsymbol{e},$$

where $\boldsymbol{m}_X = [m_{Xg} \ m_{Xb}]'$, $\boldsymbol{e} = [e_g \ e_b]'$ and $\boldsymbol{\Sigma} = \begin{bmatrix} \sigma_g^2 & \rho\sigma_g\sigma_b \\ \rho\sigma_g\sigma_b & \sigma_b^2 \end{bmatrix}$. The first-order condition (the gradient of U_X must be equal to zero) implies $-r\boldsymbol{p} + \boldsymbol{m}_X - \gamma\boldsymbol{\Sigma}\boldsymbol{w}_X + \delta_X\boldsymbol{e} = \boldsymbol{0}$, so that agents' relative demands satisfy

$$\boldsymbol{w}_X = \gamma^{-1}\boldsymbol{\Sigma}^{-1}(\boldsymbol{m}_X + \delta_X\boldsymbol{e} - r\boldsymbol{p}). \tag{4.1}$$

As δ_X increases in magnitude, the agent will progressively grant more importance to the ESG score $\boldsymbol{e}$ than to the expected payoff $\boldsymbol{m}_X$. Also, all other things equal, the magnitude of demand decreases with risk aversion and payoff volatility. Note that the demand can very well be negative. Also, if the

---

[3]Strictly speaking, the w_X are *not* portfolio weights.

correlation between payoffs (ρ) is zero, the expression simplifies to

$$w_{Xy} = \frac{m_{Xy} + \delta_X e_y - r p_y}{\gamma \sigma_y^2}, \quad X \in \{G, B\}, \quad y \in \{g, b\}.$$
(4.2)

There are three parts in the above formula. The first can be interpreted as the agent-specific attractiveness of the asset $m_{Xy} + \delta_X e_y$, which has two components: payoff and sustainability. The second is the negative impact of the price (demand decreases with price). The third (denominator) is risk. The assumption that $\rho = 0$ is quite strong, as it implies that both assets are priced independently.

Equilibrium. The price-weighted demands must satisfy the market clearing condition (demand equals supply). The aggregate demand for one asset y is simply $(q_{Gy} + q_{By}) p_y$, i.e., the price times the total holdings (in shares). In our simple setting, we assume a vector $\boldsymbol{s} = [s_g \ s_b]'$ of supply for assets, thus

$$\underbrace{\mathrm{diag}(W_G \boldsymbol{w}_G + W_B \boldsymbol{w}_B) \boldsymbol{p}}_{\text{total demand for asset}} = \boldsymbol{s}, \qquad (4.3)$$

where $\mathrm{diag}(\boldsymbol{v})$ takes a vector $\boldsymbol{v}$ as argument and yields a diagonal matrix with $\boldsymbol{v}$ as diagonal values. The above equation highlights the importance of each agent's relative weight on the market, which is given by the ratio of their wealth to total wealth $W = W_G + W_B$. To further ease the analysis, we posit that the correlation between payoffs is zero. Consequently, plugging the demands in Equation (4.2), the equation for asset y translates to

$$- \overbrace{[r(W_G + W_B)]}^{\text{riskless alternative}} p_y^2 + \overbrace{[W_G(m_{Gy} + \delta_G e_y) + W_B(m_{By} + \delta_B e_y)]}^{A_y = \text{total weighted attractiveness}} p_y$$

$$- \underbrace{\gamma \sigma_y^2 s_y}_{\text{risk/supply}} = 0,$$

where we have singled out the total attractiveness of the asset, which we write A_y. The equation is quadratic in the price p_y

because of the way the w_Y are defined. In many papers, the market clearing Equation (4.3) leads to linear forms. Under the parametric condition

$$(\boldsymbol{C}) \quad A_y^2 - 4\gamma\sigma_y^2 s_y rW \geq 0, \qquad (4.4)$$

the positive price of asset y is

$$p_y = \frac{\sqrt{A_y^2 - 4\gamma\sigma_y^2 s_y rW} + A_y}{2rW}. \qquad (4.5)$$

Intuitively, the price of an asset is increasing in attractiveness, and decreasing with risk and supply.

4.1.2 Numerical example

Now, let us test some parametric configurations of the model so as to reveal several key theoretical predictions. We make the following simplifications:

- We normalize wealths. $W_B = 1$, so that the Brown investor is the benchmark. We can take several values for W_G. For $W_G = 0.5$, the Green investor represents one third of the market and for $W_G = 2$, two-thirds of the market.
- For ease of interpretation, we fix the ESG scores to $e_g = 1$ (green) and $e_b = 0$ (brown). Also, for both assets, $\sigma_y = 0.2$ ($\sigma_y^2 = 0.04$) and $s_y = 1$ (unit supply).
- The taste for sustainability of the Green investor is set to $\delta_G = 0.15$. On the other hand, the Brown investor does not care about ESG and $\delta_B = 0$.
- Finally, we assume that $r = 0.02$. r has two important effects on prices. First, it plays a scaling role in condition $(\boldsymbol{C})$: large values of r may lead to a violation of the condition. Second, it normalizes price values in Equation (4.5), so that, at a first-order approximation, prices are inversely proportional to r.

We consider two alternative versions of agent beliefs:

1. **Extreme polarization**. The Green investor is purely driven by ESG concerns and is agnostic with respect to returns, so that $m_G = 0$. The Brown investor, in contrast, and as is commonly accepted (see Bolton and Kacperczyk (2020, 2021)), expects incremental returns for the brown asset (akin to a *carbon premium*). We model that by assuming that m_B is negatively linked to e: $m_{Bg} = 0$ (zero expected payoff from the green asset), $m_{Bb} = 0.1$ (i.e., a 10% expected payoff from the brown asset).

2. **Moderate diversity in tastes**. In this case, agents have less marked preferences. The Green investor is now interested in payoffs (in addition to ESG). Thus, $m_{Gg} = m_{Gb} = 0.1$, which means that both assets are expected to have the same average payoff. The Brown agent agrees with the Green agent on these parameters, so that $m_{Bg} = m_{Bb} = 0.1$. The only difference between the two agents is therefore in the ESG preferences, which remain the same as under extreme polarization.

In Figure 4.1, we plot the corresponding prices as functions of the wealth of the Green agent. Naturally, because the Green agent favors the ESG criterion, the price of the green asset increases when sustainable demand increases. For the brown asset, however, it is the opposite in the left panel, because the raw demand of the Green agent is negative (see Equation (4.2)). In the right panel, the Green agent has a mildly positive demand and the price of the brown asset converges (as it should) to $m_{Gb}/r = 5$. The limiting value for the green asset is $(m_{Gg} + \delta_G e_g)/r = 12.5$.

Note that in the left panel, when Green demand is too small (on the left), green prices are not defined. Reversely, when Green demand is too large, brown prices are not valued (right part of the graph). When tastes are less diverse (right panel), these issues vanish.

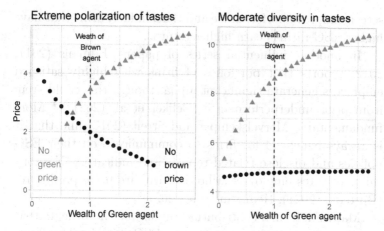

Figure 4.1 **Theoretical predictions**. We plot the price of two assets (defined in Equation (4.5)) as a function of the wealth of the Green investor (the wealth of the Brown investor is kept constant at unit value). Prices for the green asset ($e = 1$) are shown with triangles, while those for the brown asset ($e = 0$) are shown with circles. For some values of the x-axis, prices may not be defined in the left plot.

4.2   SRI IMPROVES PERFORMANCE

A large number of published work comes to the conclusion that sustainable investments are more profitable than aggregate benchmarks or even unethical portfolios. In fact, according to the survey Friede et al. (2015), 90% of papers report a positive relationship between performance and the propensity to tilt portfolios towards ESG stocks. As early as in Klassen and McLaughlin (1996), it is found that environmentally friendly corporate management is rewarded by positive returns. In another seminal article, Gompers et al. (2003) show that corporate governance is a strong (positive) driver of returns in the cross-section. Prior to that, Core et al. (1999) also concluded that weak governance is detrimental to financial performance. In a related study, Auer (2016) find that governance-related screens improve performance and that, in fact, firms that have higher ESG ratings earn higher returns.

In another influential series of papers, Edmans (2011, 2012) reports that portfolios of firms with highly satisfied employees generate significant alpha, though this could stem from prior underperformance (Celiker et al. (2021)).[4] More fundamentally, Mervelskemper and Streit (2017) find that it is always beneficial for firms to communicate on their ESG policies and disclose related reports and indicators. Notably, this motivates employees, who, in turn, are more productive (Burbano (2021), Hedblom et al. (2021)).

Many additional contributions underline the benefits that can be extracted when resorting to ESG data in the allocation process, often by selecting those firms with the highest scores. We list a few references in chronological order below in Table 4.1.

In an attempt to unify several competing theories, Giese et al. (2019) list three channels through which ESG may positively affect performance: cash flows (ESG firms yield higher dividends), risk (ESG firms have lower tail risk) and valuation (ESG firms, via a lower cost of capital, have an increased value). In a related study, Antoncic et al. (2020) find that it is not necessarily the raw ESG ratings that matter, but also their dynamics. They show that firms that experience ESG momentum (when ESG scores increase) generate positive alpha (similarly, Conen and Hartmann (2019), Shanaev and Ghimire (2021) and Tsai and Wu (2021) show that ESG score revisions matter). Changes in ESG ratings are found to have a significant impact on subsequent returns in Glück et al. (2021). The authors conclude that an improvement in the E pillar can be seen (and can act) as a risk mitigating trigger.

Likewise, Kim and Kim (2020) report that firms that enhanced their environmental sustainability by adopting cleaner

---

[4]For a critical essay on stakeholder governance, we refer to Bebchuk and Tallarita (2020). Also, at the aggregate level, Chen et al. (2020) and Symitsi and Stamolampros (2020) find that employee sentiment is a predictor of stock returns.

Reference	Notable finding
Cremers and Nair (2005)	A portfolio that is long firms with high level of takeover vulnerability and short firms with low levels of takeover vulnerability generates an annualized abnormal return of 10 to 15% if public pension fund (blockholder) ownership is high.
Derwall et al. (2005)	Eco-efficient portfolios (consisting of firms that are more environmentally friendly) generate higher returns compared to non-eco-efficient strategies.
Kempf and Osthoff (2007)	Buying high ESG stocks and selling their low ESG counterparts yields annual returns of 8.7% on average.
Evans and Peiris (2010)	ESG is positively linked to stock returns, stock valuation and operating performance.
Gil-Bazo et al. (2010)	SRI funds perform better than their conventional counterparts, even after management fees, but the out-performance comes from specialized funds (see also Filbeck et al. (2016)). SRI vehicles from general funds underperform.
Giroud and Mueller (2011)	"*Weak governance firms have lower equity returns, worse operating performance, and lower firm value, but only in noncompetitive industries.*"
Deng et al. (2013)	In M&A deals, high ESG acquirers generate higher returns and have higher success rates.
Eccles et al. (2014)	Highly sustainable firms outperform poorly sustainable firms both in terms of stock market and accounting performance.
Cai and He (2014)	It takes time for ESG to pay out. The paper reports that profitability comes after 3 years but not before.
Matsumura et al. (2014)	On average, on a 2006-2008 panel of 256 firms, for every additional thousand metric tons of carbon emissions, market capitalization decreases by $212,000.
Dimson et al. (2015)	According to the article, firms that commit to successful ESG engagements benefit from positive abnormal returns.
Krüger (2015)	Investors respond strongly negatively to negative CSR events, thereby sanctioning bad firms with negative returns. Also, stocks that face severe ESG controversies significantly underperform their benchmarks (de Franco (2020)). However, investors may be tempted to overreact to these ESG controversies (Cui and Docherty (2020)).
Nagy et al. (2016)	ESG portfolios have superior alpha, compared to a global (MSCI World) benchmark.
Verheyden et al. (2016)	ESG screens improve risk-adjusted returns.
Price and Sun (2017)	CSR is positively rewarded and CSI is penalized, but the effects of CSI last longer.
Lending et al. (2018)	Sustainable firms with small boards face lower odds of data breaches, which is beneficial for market returns.
Khan (2019)	The authors report: "*In the cross-section, forward stock returns increased monotonically across governance and ESG quartiles.*"
Kumar et al. (2019)	Firms with higher climate change exposure experience lower subsequent returns.
Li et al. (2019)	Best CSR firms earn positive abnormal returns and are more likely to have positive earnings surprises.

Table 4.1 Contributions that conclude to a positive relationship between ESG and financial performance.

Reference	Notable finding
Awaysheh et al. (2020)	Best-in-class firms (in terms of ESG scores) outperform their industry peers.
Ravina and Hentati Kaffel (2020)	The Green-minus-Carbon factor is rewarded in Europe. In addition, it helps explain the cross-section of stocks by augmenting the 5-factor model proposed by Fama and French (2015).
Serafeim (2020)	The article shows that to make ESG data profitable, it is useful to resort to public sentiment about firms' sustainability performance.
Madhavan et al. (2021)	The paper analyzes factor loadings of ESG-tilted funds. They show that they are more exposed to the quality and momentum factors and that they also have high alphas.
Khajenouri and Schmidt (2020)	The authors compare conventional equity indices with their sustainable screened counterparts. The authors find that green indices outperform with respect to Sharpe ratios.
Guest and Nerino (2020)	When a firm experiences a downgrade in governance rank from the Institutional Shareholder Services, negative returns follow.
Naffa and Fain (2020)	Nine ESG-tilted portfolios are built on sustainable themes. All of them yielded positive and significant alpha, even after transaction costs.
Abate et al. (2021)	Mutual funds that invest in high ESG stocks perform better.
Wendt et al. (2021)	A portfolio that is long low emission firms and short high emission firms earns an annual alpha of 3.6% (equally-weighted portfolio) or 5.9% (value-weighted portfolio).
Chu et al. (2021)	An aggregate ESG index significantly predicts the aggregate market return (with a positive coefficient).
Xu et al. (2021)	Firms with high climate change exposure (measured via the Palmer Drought Severity Index) experience lower future profitability
Geczy and Guerard (2021)	"*Environmental ratings interact with those forecasted returns and produce excess returns both unconditionally and conditionally*" and high ESG stocks earn higher returns than low ESG stocks.

Table 4.1 (continued) Contributions that conclude to a positive relationship between ESG and financial performance.

production practices earn significant alpha. On the other end of the spectrum, Gloßner (2021) reports that a portfolio of firms with ESG *incidents* (e.g., scandals, accidents, etc.) suffers from negative alpha. This may come from analysts downgrading their earning forecasts for these firms (Derrien et al. (2021)). Waddock and Graves (1997) even document a positive feedback loop: positive financial performance fuels ESG behaviors which in turn generate higher profitability (this optimistic finding was however subsequently challenged in the replication study by Zhao and Murrell (2016)).

Finally, Bennani et al. (2018) find that ESG has performed well *recently*. They document that all three pillars yield positive returns for long-short portfolios, but for different reasons: for E, the performance is driven by higher returns of the long leg, while for S and G, it is driven by the poor performance of the short legs.

4.3   SRI DOES NOT IMPACT PERFORMANCE

Theoretically, restricting the investment universe should be detrimental to the profitability of screened portfolios (see next subsection). One reason for this is that voluntarily omitting assets may increase the odds of missing fruitful opportunities. In fact, the reverse argument works as well: screening can also participate to exclude assets that will perform badly. All in all, these two effects compete and the aggregate outcome is unclear. Consequently, it is not surprising that many studies contend that SR investing does not really hurt financial performance (while also not improving it).

In a rebuke to Gompers et al. (2003), Johnson et al. (2009) report that screening firms on governance ratings does not yield more profitable portfolios. Similarly, Core et al. (2006), Post and Byron (2015) and Amihud et al. (2017) do not find any causal relationship between weak governance, gender balanced boards, or staggered boards, and lower market returns. At the aggregate level, the studies of Hamilton et al. (1993), Guerard Jr (1997),Statman (2000), Bauer et al.

(2005), (Bello (2005), Dolvin et al. (2019), Plagge and Grim (2020), Wee et al. (2020), Curtis et al. (2021) (on mutual funds) and Sharma et al. (2021), as well as the survey Burghof and Gehrung (2021) compare ethical funds to conventional ones and do not find significant differences in performance metrics (notably, average returns, Sharpe ratio and diversification measures). In the same vein, Capelle-Blancard and Petit (2019) report that markets react insignificantly both to positive and negative ESG news.

Furthermore, responsible screening processes in portfolio construction do not deteriorate performance (see Basso and Funari (2014), Humphrey et al. (2012), Humphrey and Tan (2014), Blankenberg and Gottschalk (2018) and Chava et al. (2021)), but they do not improve it either (Gibson et al. (2020), Pyles (2020) and Görgen et al. (2021). Sautner et al. (2021b) report a zero unconditional risk premium associated to climate change risk. In some sectors (e.g., banking), there is no significant link between CSR and CFP (Soana (2011) – though Bătae et al. (2021) do find some links depending on the components of ESG).

Likewise, market neutral portfolios built on ESG metrics are neither beneficial nor detrimental to profits (Breedt et al. (2019), Dai and Meyer-Brauns (2021)), or, equivalently, have insignificant average returns (Kaiser (2020b)). Furthermore, factor models seem to confirm that ESG is not priced (Xiao et al. (2017)), or add no incremental value, compared to traditional factors (Naffa and Fain (2021), Husse and Pippo (2021)). In Bruno et al. (2021) acknowledge that the performance of ESG leaders is appealing, unless other traditional asset pricing factors are taken into account. Indeed, the alpha of ESG portfolios disappears when controlling for the size, value, momentum, low volatility, profitability and investment factors. Overall, the meta study Kim (2019) confirms that *"SRI performance is not different from conventional investments"*. Knoll (2002) provides a theoretical argument to explain this lack of impact: the long-run demand curves for the securities of individual firms are not very steep, hence ethical screening does not shift prices much in the long run.

Baron et al. (2011) estimate a broad economic model in which firms have to deal with three markets: the financial market that prices their values, the customer market that purchases their products and a market of social pressure that incentivizes them to pursue sustainable policies. They find that CFP is uncorrelated with CSP and negatively correlated with social pressure.

Finally, we single out a few additional studies. The early meta-analysis of Arlow and Gannon (1982) finds little evidence that links social responsiveness to economic performance. Ng and Zheng (2018) document that green energy firms have similar performance compared to non-green energy firms, while Halbritter and Dorfleitner (2015) show that high ESG minus low ESG portfolios do not reveal significantly positive returns. Cheung (2011) finds that stocks that are suddenly included or removed from the Dow Jones Sustainability World Index do not experience any particular shift in average return or risk. Via Fama and MacBeth (1973) regressions, Timár et al. (2021) also does not find any significant pricing ability of ESG scores. This is also true when accounting for errors-in-variables (Auer (2021)). We sum up this section with one of the conclusions of Schröder (2004) (p. 130): "*socially screened assets seem to have no clear disadvantage concerning their performance compared to conventional assets*".

4.4   SRI IS FINANCIALLY DETRIMENTAL

At the other end of the spectrum, several studies argue that unethical firms tend to experience higher profitability. One such controversial article is that of Fabozzi et al. (2008), which shows that sin stocks (belonging to the adult services, alcohol, biotech, defense, gaming and tobacco industries) produce a combined annual return of 19%, outperforming any reasonable benchmark.[5] (Hong and Kacperczyk (2009) also find outperformance, but to a lesser extent, while Dimson et al. (2020b) do confirm superior performance of sin stocks over the long run, since 1900). Likewise, Bolton and Kacperczyk (2020, 2021) and Busch et al. (2020) find that firms with higher total CO_2 emissions (and changes in emissions) earn higher returns. The former authors conclude: "*investors are already demanding compensation for their exposure to carbon emission risk*". Similarly, Hsu et al. (2021) document that polluting firms experience higher average returns. Moreover, Delmas et al. (2015) also report that improving corporate environmental performance reduces short-term financial performance. In the same vein, Trinks and Scholtens (2017) observe that investing in controversial stocks in many cases results in superior risk-adjusted returns.

Conversely, firms that make significant environmental efforts experience lower returns (Fisher-Vanden and Thorburn (2011)), just as do *impact* firms (Bernal et al. (2021)). One explanation is that reducing the carbon footprint or any source of pollution is too costly and outweighs the potential benefits (Hart and Ahuja (1996)). A second explanation is under-diversification. On the governance side, Bøhren and Staubo (2016) report that firms with mandatory gender parity on their boards experience lower returns (though a broader examination of the literature leads to a more nuanced conclusion, see Post and Byron (2015)).

At the aggregate level, in their study on university endowment funds, Aragon et al. (2020) find that funds which enforce sustainable policies have greater volatility and the authors attribute underperformance to higher divestment costs and inefficient diversification. Similarly, Anson et al. (2020) find that sustainable funds have, on average, lower alphas

---

[5]In a follow-up paper, Blitz and Fabozzi (2017), it is nonetheless found that a five-factor model is able to explain this sin stock anomaly. Recently, Blitz and Swinkels (2021a) also find little evidence of a sin premium, but warn that it could materialize in the future, when sin firms will be asked higher costs of capital.

compared to unconstrained funds (see also Liang et al. (2021), who show that, nonetheless, ethical hedge funds attract more flows). Bofinger et al. (2021) find that active ESG funds face the risk of lowering their investment skills because of what the authors call the *sustainable trap* (the push towards ESG increases the risk of mispricing). Similar findings are outlined in Renneboog et al. (2008a) and El Ghoul et al. (2021), though the underperformance in returns sometimes translates to statistically insignificantly lower Sharpe ratios. Finally, focusing on the debate on mandatory reductions of greenhouse gases in the U.S., Hsu and Wang (2013) report that markets react positively to negative ESG news at the company level. Conen and Hartmann (2019) document a reverse effect: *"markets react to ESG improvements of top ranked firms with negative abnormal returns."*

From an asset pricing perspective, and according to risk factor analysis, investors should be rewarded for the risk they take when investing in stocks that are exposed to ESG externalities. This is one of the theoretical findings of Pastor et al. (2021b): the capital asset pricing model (CAPM) alpha of stocks should be negatively proportional to their ESG scores. And indeed, Chen and Scholtens (2018) argue that both active and passive SRI funds have negative alphas. Empirically, Brammer et al. (2006), Lee and Faff (2009), Becchetti et al. (2018), Lioui et al. (2018), Lioui (2018a,b), Ciciretti et al. (2020), Adriaan Boermans and Galema (2020), Hübel and Scholz (2020), Alessi et al. (2020), Alessi et al. (2021) and Cakici and Zaremba (2021) all find that the rewarded ESG factors go long irresponsible firms and short responsible ones. The latter documents that a significant part of the premium comes from small firms (which disclose typically less). Similarly, a *boycott* portfolio, long sin stocks and short non-sin stock earns a monthly return of 1.33% on average (Luo and Balvers (2017)). In Kanuri (2020), it is found that in the long run, conventional funds outperform ESG funds (in terms of average returns and Sharpe ratio), even though the latter sometimes fare better.

Also, from an optimization standpoint, using screening processes reduces the investment set. By construction, this shifts the efficient frontier towards a smaller enveloppe, which is, by definition, suboptimal – at least in-sample. This implies the opportunity cost of renouncing to potential profitable assets. For a theoretical discussion on this subject, we refer to Pedersen et al. (2021). In particular, Chang and Witte (2010) argue that ESG investing shrinks both average returns and Sharpe ratios, compared to unscreened benchmarks.

Finally, in an original contribution, Jørgensen and Plovst (2021) analyze the hedging cost of sustainability. They measure the price of an insurance derivative that would protect an ESG investor against the underperformance of a green fund versus a conventional fund. The authors find that the incurred cost lies between –0.5% and –3% in terms of annual returns.

4.5 IT DEPENDS...

Not surprisingly, SRI does not deliver performance unconditionally, if and when it does. There are in fact many drivers of this performance and the profitability of ESG strategy can depend on several factors, which we list below. The meta-analysis Hang et al. (2018) is a valuable resource on this topic.

- First of all, not all **dimensions** of ESG are equal. Some studies find that governance screens work well (Gompers et al. (2003), Bebchuk et al. (2009), Auer (2016), Bruder et al. (2019) and Lee et al. (2020), but environmental screens do not (Auer (2016), Alareeni and Hamdan (2020)). In fact, even within ratings, some provisions, or subcategories are more impactful than others (see Bebchuk et al. (2009) and Becchetti et al. (2018)). In a similar vein, Ziegler et al. (2007) finds that E is improving returns, while S is deteriorating them. Likewise, Galema et al. (2008) and Jayachandran et al. (2013) show that some ratings can be beneficial (diversity, environment and product), while others cannot. Within each branch of ESG, some component may also prove more relevant than others. Jacobs et al. (2010) find that philanthropic gifts for environmental causes are associated with significant positive market reaction, while it is the opposite for voluntary emission reduction.

Environmental commitments are not all equal: King and Lenox (2002) find that prevention leads to financial gain, but not pollution reduction. Based on KLD data, Geczy et al. (2019) show that Human Rights, and Diversity criteria contribute to enhancing portfolio returns. Filbeck et al. (2019) find that G is beneficial, E is detrimental and S is not impactful. Giese et al. (2021) find that pillars can matter at different horizons (short for Governance, long for Social and Environmental). Even inside pillars, variables may have mitigating effects. Naranjo Tuesta et al. (2020) find that different types of policies on different types of emissions have contradicting effects on financial performance. According to Tsai and Wu (2021), enhancing the environmental score leads to higher relative returns in crisis periods.

Moreover, depending on which fields are considered, results differ. Aswani et al. (2021) underline that academics often use the raw values of emissions (in metric tons of CO_2), while it makes more sense to consider emission intensities, whereby emissions are scaled by some proxy of firm size (e.g., sales) – which may considerably alter conclusions. Furthermore, depending on the industry (see also below), some ESG fields may be considered as *financially material* (i.e., expected to have an impact on finances), while other are not. The study Grewal et al. (2021) shows that materiality is a strong driver of price informativeness.

Beyond pure ESG fields, the propensity of firms to simply disclose ESG related data is also useful in the allocation process (D'Apice et al. (2021)). Finally, in De Franco et al. (2021), the authors warn that the ESG dimensions are not the only ones in the field. According to them, the sustainable development goals (SDGs) allow to elaborate alternative metrics that capture other facets of sustainability. For an intelligible introduction on SDGs and their relationships with finance, we recommend the overview of Zhan and Santos-Paulino (2021).

- A second important factor is **geography**. Many articles document contradicting results when switching from one market to another. Often, ESG strategies are shown to perform for one zone (e.g., US, Europe, or Asia), but not all. We refer for instance to Cortez et al. (2012), Von Arx and Ziegler (2014), Post and Byron (2015), Bruder et al. (2019), Cheema-Fox et al. (2019), Matallín-Sáez et al. (2019), de Franco (2020), Edmans et al. (2020), Griffin et al. (2021), Amon et al. (2021), Murata and Hamori (2021). and Giese et al. (2021) In addition, in Chakrabarti and Sen (2020), it is shown that global indices outperform the market, but regional indices do not. In some papers, however, it is shown that there is no link between sustainability and performance, no matter the geographical zone (see Auer and Schuhmacher (2016)). Lastly, there are so many country-specific studies that it is impossible to cite them exhaustively.

- The third important factor is **chronology**. Just like all factor tilts in the world, the ESG *style* has its moments. The perception of ESG has shifted through time. Ioannou and Serafeim (2015) find that analysts' recommendations *used to be* pessimistic toward corporate SR policies. Political changes also impact expectations toward green firms and generate shocks to returns (Ramelli et al. (2021)). We refer to Bauer et al. (2005), Barnett and Salomon (2006), Statman (2006), Climent and Soriano (2011), Bebchuk et al. (2013), De and Clayman (2015), Belghitar et al. (2017), Ibikunle and Steffen (2017), In et al. (2017), Bennani et al. (2018), Mutlu et al. (2018), Ciciretti et al. (2020), Li et al. (2019), Vojtko and Padysak (2019), Kanuri (2020), Bansal et al.

(2021), Faccini et al. (2021) and Tsai and Wu (2021) for contributions that document the time-varying nature of sustainable investments. Barnett (2017) reports that the price of climate risk (evaluated via temperatures) is more significant after 1996, compared to the full sample period of 1948-2019.

Several of the aforementioned studies underline that the profitability of SRI was highest in the most recent periods. The hype surrounding ESG investing (and the related demand) may have pushed the prices up. For instance, it is shown in Azar et al. (2021) that the commitment of global asset managers with respect to sustainability is relatively novel and significant. Relatedly, Bansal et al. (2021) find that SRI performs well in good times, but badly during economic downturns and crises; Nofsinger and Varma (2014) come to the opposite conclusion.

Lioui and Tarelli (2021) disentangle time-series and cross-sectional ESG factors. They document a significant time variation in the alpha of both types of factors, conditional on firm characteristics.

The contribution of Pastor et al. (2021a) points out that the recent outperformance of sustainable firms is entirely due to aggregate concern over the environment. The authors use the Media Climate Change Concerns (MCCC) index of Ardia et al. (2020) and prove that the long-short green factor sees its performance vanish once it has been controlled for shocks in the MCCC index.

- A fourth dimension is **industry**. For instance, Herremans et al. (1993), Russo and Fouts (1997), Semenova and Hassel (2008), Hoepner and Yu (2017a), Giroud and Mueller (2011), Lee et al. (2011) (oil and gas), Jo and Na (2012), Auer and Schuhmacher (2016), De Haan and Vlahu (2016), Feng et al. (2017), Bertolotti and Kent (2019), Alessandrini and Jondeau (2020), Torre et al. (2020), Abdi et al. (2021) (airline), Giese et al. (2021),

Kuo et al. (2021) (airlines) and Okafor et al. (2021) find that the impact of ESG depends on the sector of firms. In contrast, Statman and Glushkov (2009) argues that for SRI to reach its full potential, it must only rely on ESG ratings and not on industry screening. Giese et al. (2021) argue that the weighting of ESG pillars should be sector-specific. Typically, industries are not hit uniformly by rising temperature (see Shaw et al. (2021)). For an analysis on the energy sector, we refer to Brzeszczynski et al. (2019).

- Finally, a fifth mitigating effect is firm **ownership**. Nekhili et al. (2017) and Abeysekera and Fernando (2020) indicate that family involvement and ownership is also likely to impact the relationship between CSR and financial performance. Brøgger and Kronies (2020) find that the positivity (and significance) of the ESG factor exists within firms that are owned by unconstrained investors (e.g., mutual and hedge funds). Relatedly, Cheema-Fox et al. (2019) find that the performance of portfolios is strongly linked to institutional investor flows.[6]

Some researchers (e.g., Barnett and Salomon (2006), Brammer and Millington (2008), Fernando et al. (2010), Harjoto et al. (2017) and Gerged et al. (2021)) manage to reconcile seemingly contradicting results by showing that the relationship between CSR and performance is not linear: very good and very bad corporations experience abnormal returns, while those in the bulk of the ESG distribution perform differently. Xie et al. (2019) show that the optimal level of disclosure is in the middle of the distribution and very low or very high disclosure leads to lower performance. de la Fuente et al. (2021) also document an inverted U-shape. The screening intensity and the type of screens seems to play a role as well

---

[6]Institutional ownership affects the propensity of firms to indulge in ESG policies, see, e.g., Martínez-Ferrero and Lozano (2021) in the case of developing countries.

(Capelle-Blancard and Monjon (2014)), if only because it impacts the diversification of the portfolio (Jin (2020)). Fairhurst and Greene (2020) also document a non-monotonic impact of ESG scores: extreme CSR policies appear to be harmful, at least on the takeover market. Non-linear patterns are moreover documented in Huang and Hilary (2018) for governance proxies. Asymmetric preferences of investors (who are indifferent to best-in-class, but penalize worst-in-class firms with negative E and G scores) are revealed in Nofsinger et al. (2019). One other route to explain diverging results in the field is to argue that models are misspecified, e.g., when important independent variables are omitted (see McWilliams and Siegel (2000)). The way ESG criteria are integrated in the portfolio design can also matter. In their study on the Australian market, Fan and Michalski (2020) find that simple ESG sorts have disappointing performance, but combined with other factors, like quality or momentum, boosts their returns.

The way and reason why firms disclose CSR actions is also likely to matter. Bams et al. (2021) separate three dimensions in CSR performance and disclosure. First, strategic CSR refers to genuine sustainability for the sake of sustainability. Second, CSR as insurance is a more passive approach to sustainability through which boards and firms *"conform to the institutional pressure for CSR by providing a minimal level of CSR to mitigate risks and maintain their licence to operate"*. Finally, there is greenwashing. Bams et al. (2021) find that firms which choose the former outperform the others in both realised social and financial dimensions.

We end this subsection with four references. Barnett and Salomon (2006) contend that risk-adjusted returns of ESG strategies depend non-linearly in screening intensities. López-Arceiz et al. (2018) argue that the profitability of SR funds is strongly impacted by the cultural environment in which the fund operates. Dorfleitner et al. (2020) find that the weighting scheme of portfolios can matter, as well as the size of companies. Finally, Ardia et al. (2020) (following Pastor et al. (2021b)) find that the profitability of green minus brown portfolios depends on the aggregate concern with respect to climate threats.

4.6 CSR AND RISK

The question of whether SRI is an efficient way to hedge risk remains open. This has critical implications for institutions such as pension funds (Sautner and Starks (2021)) in the case of downside risk. Becchetti et al. (2015) find that SR funds performed less badly during the 2007-2008 financial crisis, compared to conventional funds. De and Clayman (2015) and Hoepner et al. (2021) document a negative relationship between ESG ratings and risk (measured by stock volatility and downside variance). Dunbar et al. (2021) find that CSR is only a vector of risk reduction if the governance of firms seeks transparency and corporate social performance. In their study on European funds, Gonçalves et al. (2021) reveal that green funds outperform conventional funds in times of crises.

Finally, two rather contrarian articles, Brav and Heaton (2021) and Heaton (2021) suggest that "prudent" investors might want to invest in brown assets in order to hedge against the likelihood that the transition to a greener economy fails to materialize. They highlight the probabilities of scenarii in which brown assets would outperform their green counterparts.

The COVID-19 pandemic yielded a thread of event studies that aim at understanding if sustainability mitigated risk. We list a few such contributions below. Singh (2020) finds that the ESG factor performed well during the COVID-19 market crash (see Mahmoud and Meyer (2021) for an in-depth analysis of the drivers of ESG preferences posterior to this drawdown). Similarly, Pástor and Vorsatz (2020) and Omura et al. (2020) report that sustainable funds outperformed conventional ones during the market meltdown. Akhtar et al. (2021) document a positive effect of gender diversity on the abnormal performance of US stocks during the period between March and April 2020. In Xiong (2021), firms with low ESG risk are found to be outperforming those with high ESG risk. Likewise, in their study on Indian stocks, Arora et al. (2021) reveal that sustainable firms fared better than brown ones during the pandemic. Similar results are obtained on European data in Pizzutilo (2021). However, Folger-Laronde et al. (2020), Demers et al. (2021), Mahmoud and Meyer (2020), Capelle-Blancard et al. (2021), Chiappini et al. (2021), Gianfrate et al. (2021), Pavlova and de Boyrie (2021) and Yousaf et al. (2021) see little or no hedging power of ESG-driven funds during the pandemic.[7] Glossner et al. (2021) find no evidence that investors shifted toward ESG firms during the COVID-19 crisis and Singh (2021) finds evidence that investors shifted from ESG *equities* to ESG *bonds*. Finally, ESG returns may also be contingent on investor sentiment (Azevedo et al. (2021)).

Below, we list further contributions that conclude that ESG is positively, negatively, or weakly linked to risk.

- **ESG reduces risk**. CSR engagement reduces risk in controversial industries (Jo and Na (2012)). Social irresponsibility is linked to higher risks (Oikonomou et al. (2012)). Board quality reduces many types of risk (Kaiser (2020a)). Hedge funds can benefit from risk-mitigating properties of ESG investments (Duanmu et al. (2021)). A few articles favorably compare traditional indices to their green counterparts during downturns (Becchetti et al. (2015), Gonçalves et al. (2021), Ouchen (2021)). Other contributions include: McGuire et al. (1988), El Ghoul et al. (2011), Humphrey et al. (2012) and Kong et al. (2020) (idiosyncratic risk), Harjoto et al. (2017), Bernile et al. (2018), Monti et al. (2019), Gibson et al. (2020), Lopez de Silanes et al. (2020), Karwowski and Raulinajtys-Grzybek (2021). ESG also seems negatively related to *tail* risk (Bax et al. (2021), Xiong (2021)), to stock crash risk (Wu and You (2021) find negative correlation with the quality of green

---

[7]Yousaf et al. (2021) find that green *bonds* are however useful.

patents, but not their quantity), and to default risk (Aslan et al. (2021)).

- **ESG is riskier.** Bianchi et al. (2010) document a negative effect of ESG: green funds have higher betas than conventional funds during recessions. Ng and Rezaee (2020) find that ESG performance is positively linked to idiosyncratic volatility. Fiordelisi et al. (2021) find that ESG ETFs perform worse than the market during economic downturns. Similar conclusions are obtained by Bansal et al. (2021).

- **Link is not clear.** Just like for pure performance, risk is not unequivocally linked to ESG. CSR is weakly linked to risk (Oikonomou et al. (2012)).

- **Downside risk.** On this topic, we refer to Monti et al. (2019), Shafer and Szado (2020), Nofsinger et al. (2019) and Richey (2020). The latter shows that sin stocks are defensive in bad times. For a focus on US banks, we recommend the contributions of Tasnia et al. (2021) and Chaudhry et al. (2021).

- **Implied volatility.** Firms that belong to industries with higher ESG-sales dynamism are associated with lower implied volatilities (Patel et al. (2020)).

4.7 ESG AND OTHER FINANCIAL METRICS

Beyond pure stock-market profitability, ESG seems to be favorably related to the metrics listed in Table 4.2. With respect to valuation and cost of equity, let us briefly recall that, according to Gordon's dividend model, the firm value is equal to $V_0/(c-g)$, where c is the cost of capital (e.g., weighted average cost of capital (WACC)) and g is the growth rate of cash flows (or dividends). ESG issues can impact both channels. Firm value can increase if g increases, or if c decreases, e.g., because investors perceive lower risk. Derrien et al. (2021) show that with respect to ESG incidents, it is the first channel that matters most.

Performance metric	References
Raw valuation	Hillman and Keim (2001), Konar and Cohen (2001), Clarkson et al. (2004), Bebchuk and Cohen (2005), Hill et al. (2007), Hong and Kacperczyk (2009), Ammann et al. (2011), Guenster et al. (2011), Cai et al. (2012) (in sin industries), Lourenço et al. (2012), Servaes and Tamayo (2013), Lund and Schonlau (2016), Amihud et al. (2017), Bajic and Yurtoglu (2018), Tsukioka (2018), De Villiers et al. (2020), Johnson et al. (2020), Ahsan et al. (2021) (in China). Cremers et al. (2017) document an absence of impact of staggered boards. This topic is reviewed in Gerard (2018).
Tobin's q	Mehran (1995), King and Lenox (2001), King and Lenox (2002), Jiao (2010), Jo and Harjoto (2011), Cai et al. (2012) (in sin industries), Lioui and Sharma (2012) (negative relationship), Cremers and Ferrell (2014), Gregory and Whittaker (2013), Lee et al. (2015) (negatively impacted by CO_2 emissions, but positively by environmental R&D), Ferrell et al. (2016), Kang et al. (2016), Velte (2017) (no impact), Hasan et al. (2018), Tsukioka (2018), Radhouane et al. (2018), Ioannou and Serafeim (2019), Alareeni and Hamdan (2020), Gantchev and Giannetti (2020), Gerged et al. (2021) (in the Gulf countries).
Return on assets, and return on equity	Herremans et al. (1993), Mehran (1995), Hart and Ahuja (1996), Russo and Fouts (1997), King and Lenox (2002), Simpson and Kohers (2002), Semenova and Hassel (2008), Peiris and Evans (2010), Guenster et al. (2011), Lioui and Sharma (2012) (negative relationship), Christiansen et al. (2016), Velte (2017), Tsukioka (2018), Yin et al. (2019), Alareeni and Hamdan (2020), Gantchev and Giannetti (2020), Gerged et al. (2021) (in the Gulf countries), Mohamed Buallay et al. (2021) (financial sector), Rossi et al. (2021).
Cost of equity	Sharfman and Fernando (2008), Dhaliwal et al. (2011), El Ghoul et al. (2011), He et al. (2013), Chava (2014), Ng and Rezaee (2015), Park and Noh (2018), Gao et al. (2021), Matsumura et al. (2020), Mariani et al. (2021) (impact on the weighted average cost of capital).
Operating performance	Guenster et al. (2011)) and earnings (Borgers et al. (2013), Velte (2019), Kim and Kim (2020), Jia and Li (2021).
Cost of debt and credit rating	Ashbaugh-Skaife et al. (2006), Attig et al. (2013), Chava (2014), Jiraporn et al. (2014), Jung et al. (2018), Caragnano et al. (2020), Kling et al. (2021), Michalski and Low (2021) and Raimo et al. (2021); and default risk (Nadaraja et al. (2020)).
Equity forecasts	Gregory et al. (2014) (with rationale of discounted cash flows), Derrien et al. (2021).
Dividend policy	Cheung et al. (2018), Benlemlih (2019), De Villiers et al. (2020) and Matos et al. (2020).
Trade credit	Cheung and Pok (2019), Xu et al. (2020).
Exit scenarios and M&A	For public firms (being acquired, going bankrupt, or going private), see Goktan et al. (2018). For M&A, see Gomes and Marsat (2018) and Caiazza et al. (2021), Arouri et al. (2019) (on uncertainty), Alexandridis et al. (2021) (on CSR culture), Jost et al. (2021) (the effect is limited), Yen and André (2019) (on emerging markets) and Reynolds and Hassett (2021) for a topical discussion.
Equity offerings	Dutordoir et al. (2018).
IPOs	Reber et al. (2021) (voluntary disclosure reduces risk post-IPO).
Employee morale and productivity	Burbano (2021), Hedblom et al. (2021).

Table 4.2 ESG and other performance metrics.

4.8   EMPIRICAL ILLUSTRATION

We close this chapter with a small empirical exercise. We compare two US equity indices: one conventional, and one ESG-
based. The conventional index is the S&P 500, arguably the
reference yardstick for US equities, both among practitioners
and scholars alike. The ESG portfolio is the iShares MSCI
USA ESG Select ETF, which performs sustainability screens
that are based on sectors, as well as on ESG scores. In December 2020, the index comprised 202 stocks, which makes it less
diversified, compared to the S&P 500.[8] The series start on January 28th, 2005, which is the inception date of the ESG index.
It is notoriously complicated to find reliable ESG data prior to
2005. In Figure 4.2, we plot the time-series of the two indices.

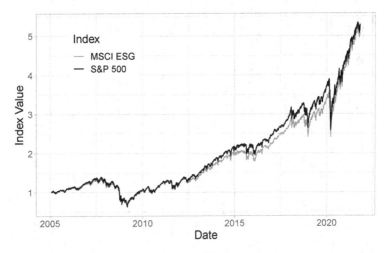

Figure 4.2 **Performance comparison**. We plot the index values (S&P 500 and iShares MSCI USA ESG Select ETF), on-
ward from January 28th, 2005 (inception date of the ESG portfolio). The series are normalized so that their initial value is
one.

---

[8]This is subject for debate because the index is capitalization-
weighted, which means that a few dozens of stocks account for a large
majority of the weights of the portfolio.

At first sight, the first-order conclusion is that there is not much difference between the two series. In the first half of the sample, the lines are hardly distinguishable. Between 2012 and 2019, the S&P 500 seems to outperform marginally, but 2020 has eroded part of this superiority (see below). In Table 4.3, we compute a few performance metrics.

Index	Return	Volatility	Ratio	Max. Drawdown	Value at Risk (5%)
MSCI ESG	0.103	0.184	0.558	-0.329	-0.017
S&P 500	0.104	0.194	0.538	-0.337	-0.018

Table 4.3 **Performance metrics**. We present the annualized compounded return, the daily volatility times $\sqrt{252}$, the ratio between the former and the latter, the maximum drawdown and the daily Value-at-Risk at the 5% level. All metrics are computed from 2005 to 2021.

All values are arguably close, and no difference in any metric would pass a test of statistical significance. The broad market index has a marginal superiority in returns, but it is bested across both risk measures. The volatility-adjusted average return is even slightly higher for the ESG index. Overall these results are most in line with those of Section 4.3: in the long run, its is hard to find evidence (in this small sample) of outperformance in one way or the other.

Researchers and practitioners often seek to determine if ESG exposure acts as a hedge in bad times. To shed some light on this question, we zoom in on two sub-periods of the sample, namely the years 2008–2009, and 2020. These are shown in Figure 4.3, where the series are scaled to start at unit value.

In 2020 (right panel), the two curves move closely together until April, which means that the sustainable tilt did not immunize the ESG portfolio against the crash.[9] To a certain extent, this is also true for the subprime crisis in 2008.

---

[9]Again, researchers disagree on this matter. Diaz et al. (2021) and Singh (2020) report some hedging benefits from ESG exposure, while Folger-Laronde et al. (2020), Demers et al. (2021) and Mahmoud and Meyer (2020) do not.

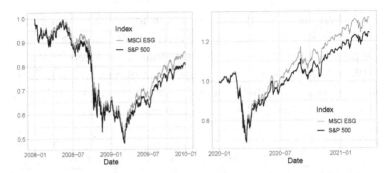

Figure 4.3 **Focus on 2008–2009 and 2020–2021**. We plot the index values, from January 1st, 2008 to December 31st, 2009 (left panel) and from January 1st, 2020 to the end of March 2021 (right panel). The series are normalized so that their initial value is 1.

However, the interesting pattern is probably revealed *after* the crises. It is in the aftermath of the crashes that differences materialize, to the benefit of the ESG index. It is as if, after being burnt by an extreme event, investors redirect flows toward more sustainable assets. This is consistent with some conclusions of Dyck et al. (2019). Nevertheless, zooming out back to Figure 4.2 reveals that this effect fades. After 2009, the S&P 500 outperformed until 2019 (see Figure 4.4), as if the appeal of ESG stocks decreases with the time span after a crash.

To illustrate the shifting relative risk of ESG portfolios, we compute the volatility ratio between the S&P 500 and the ESG index in Figure 4.5. Realized volatilities are computed as the standard deviation of the daily returns over the past 60 trading days. While the S&P 500 is more often the most volatile index, the ESG portfolio does run through some pockets of superior risk, especially between 2016 and 2019.

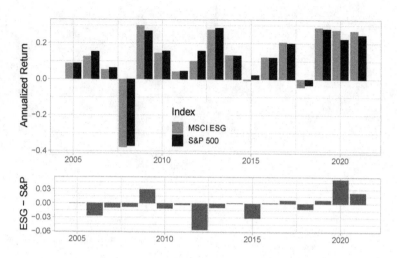

Figure 4.4 **Annual returns**. In the upper panel, we plot the average daily returns multiplied by 252, on a year-by-year basis. The lower panel shows the difference between the two indices (MSCI ESG minus S&P 500). **Note**: the values for the last year are computed on data up to October 2021.

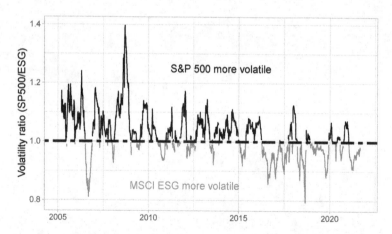

Figure 4.5 **Volatility ratio**. We plot the ratio of the S&P 500 volatility divided by the volatility of the MSCI ESG index. Realized volatilities are computed as the standard deviation of the daily returns over the past 60 trading days.

CHAPTER 5

Quantitative portfolio construction with ESG data and criteria

5.1 SIMPLE PORTFOLIO CHOICE SOLUTIONS

ESG indicators are very well suited for portfolio integration and dozens, if not hundreds of articles have been written on this topic. We begin by mentioning a few papers that are very thoroughly written and provide the reader with a detailed account of the portfolio construction process. For instance, De and Clayman (2015) give a precise description of their methodology and report many results. They find that the best solution is to eliminate stocks from the lower tail of the ESG distribution. Oikonomou et al. (2018) test several quantitative methods (mean-variance, Black-Litterman, and robust estimation) and find that they work better than simple heuristic approaches (e.g., equally-weighted portfolios). Alessandrini and Jondeau (2020) perform robustness checks across regions and sectors, but also compare best-in-class versus exclusion approaches. Lastly, Bruder et al. (2019) also look at the importance of regions and disentangle the three dimensions of ESG and their impact on returns and information ratios.

DOI: 10.1201/9781003215257-5

5.2 IMPROVED MEAN-VARIANCE ALLOCATION

One common direction that is pursued by researchers is to integrate ESG scores directly in the utility function that the investor maximizes. This makes sense because, indeed, investors value sustainability and do obtain utility outside financial gains (see Bollen (2007), Riedl and Smeets (2017), Barber et al. (2021), Dyck et al. (2019), Hartzmark and Sussman (2019)). Thus, instead of maximizing expected returns for a given level of risk, agents may seek to maximize a combination between returns and weighted-average ESG scores. This simple idea is presented in Barracchini and Addessi (2012) and Gasser et al. (2017). Some variants of these methods are (in chronological order):

- Drut (2010) proposes the maximization of a mean-variance utility, subject to a minimum level of aggregate ESG score.
- Dorfleitner and Utz (2012) introduce a utility function that is based on financial and sustainability returns of the portfolio.
- Benedetti et al. (2021) integrate views about carbon taxes within a Bayesian portfolio optimization problem.
- Branch et al. (2019) resort to optimized exclusion: they first shrink the investment set and then optimize on the remaining assets.
- Fish et al. (2019) adjust returns with ESG metrics before they run optimizations (mean-variance and hierarchical risk parity).
- Geczy et al. (2019) implement the so-called APT Mean-Variance Tracking Error at Risk optimization.
- Alessandrini and Jondeau (2020) maximize the ESG rating of the portfolio under many constraints (related to benchmark tracking error, turnover, regional and industry exposition, etc.).
- Bender et al. (2020) add carbon constraints to their objective functions and show that it is possible to reduce

the carbon emissions of portfolios without degrading the volatility of low-volatility portfolios.

- Chen and Mussalli (2020) propose to optimize the information ratio of a portfolio while integrating both alpha and ESG considerations.
- Chan et al. (2020) optimize a traditional quadratic utility function, but with constraints on factor exposures and with additional requirements on aggregate ESG and carbon scores.
- Schmidt (2020) minimizes the volatility minus the ESG score under a constraint of expected return.
- In Pedersen et al. (2021), the agent maximizes the Sharpe ratio of the portfolio for a given level of ESG score.
- The paper Geczy et al. (2021) (originally written in the early 2000s) focuses on allocations to mutual funds. The agent postulates a model for future returns and these models are subject to uncertainty, which is modelled in a Bayesian framework. The authors determine the financial cost of screening funds according to whether they are socially responsible or not. They provide some conditions under which this cost is high or low.
- Coqueret et al. (2021) propose to combine ESG attributes to other variables in the asset pricing literature (e.g., market capitalization or valuation ratios) to boost ESG-based optimization.

A very favorable feature of most of these frameworks is that they often allow closed-form solutions that are easily implementable. Henceforth, we present the framework laid out in Pedersen et al. (2021). For consistency purposes, we follow the notations of the model of Pastor et al. (2021b) (which is outlined in the online Chapter).

The agent chooses an allocation w over N assets so that future wealth $W_1 = W_0(1 + r_f + w'r)$, where r_f is the risk-free rate and r is the vector of returns of the risky assets. In addition to returns, assets are characterized by an observable ESG score g (which here has a vector form). The weighted average ESG score of the portfolio is thus $\bar{g} = w'g/(w'1)$, where here 1 denotes a vector of N ones.[1] As is discussed in Section 3.1, the investor has preferences over both pecuniary and social performance, and seeks to solve the following program:

$$\max_{w} \left\{ \underbrace{w'\mu - \frac{\gamma}{2}w'\Sigma w}_{\text{mean-variance}} + \underbrace{f\left(\frac{w'g}{w'1}\right)}_{\text{ESG tilt}}, \text{ subject to } w'1 > 0 \right\},$$

(5.1)

where μ and Σ are the mean vector and covariance matrix of the returns r, and f is the ESG preference function. As is customary, $\gamma > 0$ is the risk aversion parameter. The constraint $w'1 > 0$ implies that the portfolio has a long-only bias, i.e., that long positions strictly outweigh short ones. Under additional factor assumptions, Varmaz et al. (2021) show that it is possible to simplify this program to a plain quadratic optimization with linear constraints.

In traditional mean-variance analysis, the agent has the choice between fixing one level of average return and minimizing the volatility, or maximizing the average return of a given level of return dispersion. In the above formulation, there is one additional choice to make, and it relates to the average ESG score $\bar{g}$. The efficient frontier is defined by maximizing the expected return under constraints

$$\max \left\{ w'\mu, \text{ subject to } \begin{array}{ll} w'1 > 0, & \text{leverage constraint} \\ \sqrt{w'\Sigma w} < \sigma_{\text{target}}, & \text{volatility constraint} \\ w'g > g_{\text{target}} & \text{ESG constraint} \end{array} \right\},$$

(5.2)

or minimizing volatility under constraints

$$\min \left\{ \sqrt{w'\Sigma w}, \text{ subject to } \begin{array}{ll} w'1 > 0, & \text{leverage constraint} \\ w'\mu > \mu_{\text{target}}, & \text{return constraint} \\ w'g > g_{\text{target}} & \text{ESG constraint} \end{array} \right\}.$$

(5.3)

---

[1] ESG score is not equivalent to ESG risk, see, e.g., Gaussel and Le Saint (2020).

This implies that the frontier is a function of the target for the ESG score. This is shown in Figure 5.1, where we represent three different stylized frontiers. As the investor asks for a higher ESG score, the frontier shifts away from the unconstrained frontier and the investor must accept possibly lower returns and/or higher volatility.

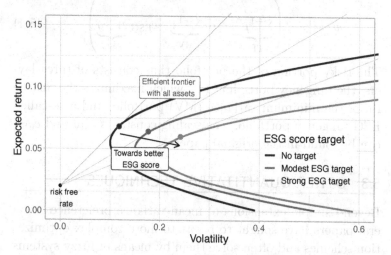

Figure 5.1 **Stylized representation of ESG constrained frontiers**. We present a diagram that depicts three efficient frontiers for various levels of targeted ESG score. The big dots at the intersection of lines and frontiers are the tangency portfolios.

In a typical mean-variance optimization, the best portfolio is the one that maximizes the Sharpe ratio (tangency portfolio). It is thus useful to characterize the maximum Sharpe ratio that can be attained for a given level of aggregate ESG score:

$$\text{SR}(\bar{g}) = \max_{w} \left\{ \frac{w'\mu}{\sqrt{w'\Sigma w}}, \text{ subject to } \begin{bmatrix} w'1 > 0 \\ w'g/(w'1) = \bar{g} \end{bmatrix} \right\}.$$

In order to solve the original program defined in Equation (5.1), the agent must choose the following level of average ESG score:

$$g^* = \max_{\bar{g}} \left\{ SR(\bar{g})^2 + 2\gamma f(\bar{g}) \right\}.$$

The general solution to Equation (5.1) is a fund separation that reads:

$$w^* = \gamma^{-1} \left(\underbrace{\Sigma^{-1}\mu}_{TP} + c_{MV} \underbrace{\Sigma^{-1}1}_{MV} + c_{ESG} \underbrace{\Sigma^{-1}g}_{ESG} \right). \qquad (5.4)$$

The risky portion of the portfolio thus consists of three layers: the tangency portfolio (TP) which maximizes the Sharpe ratio, the minimum variance (MV) portfolio, and a so-called ESG tangency portfolio. The scaling constants c_{MV} and c_{ESG} can be found in the original paper.

5.3 OTHER QUANTITATIVE TECHNIQUES

Beyond simple extensions of mean-variance preferences, several papers have sought to resort to more complex optimization schemes and often solve them by means of fuzzy systems or genetic algorithms. We refer to the following list (chronologically sorted): Bilbao-Terol et al. (2012a), Bilbao-Terol et al. (2012b), Calvo et al. (2015), Bilbao-Terol et al. (2016) and Hilario-Caballero et al. (2020). In Chen et al. (2021), the authors resort to screening methods combined to ESG score and portfolio optimization. Likewise, Alessandrini and Jondeau (2021) propose a very exhaustive optimization scheme in which the ESG score is maximized, subject to a large palette of constraints: tracking error, turnover, regional and industry weights, factor exposure, and box constraints. The only drawback is the fundamentally black-box nature of the outcome.

Other contributions go even further and rely on systematic approaches, often linked to machine learning. For instance, Lanza et al. (2020) compute brute force trees (test all combinations of ESG indicators) to derive the best possible portfolios. In a similar fashion, Margot et al. (2021) derives an ad-hoc ML algorithm that tries to link ESG ratings to returns in a non-linear fashion. Goldberg and Mouti (2019) resort to supervised learning to predict maximum drawdown and find that ESG scores help improve the forecasting accuracy of the algorithms. Sokolov et al. (2021) mix natural language processing with Black-Litterman views to build ESG-tilted portfolios.

Lastly, though it does not refer to portfolio construction, but performance attribution, we mention the work of Bolliger and Cornilly (2021). The authors detail a methodology that aims to evaluate the carbon footprint of a portfolio. The authors contend that this can largely be driven by sectorial biases and that this dimension must be taken into account.

5.4 MISCELLANEOUS TIPS, METHODS AND OTHER INTEGRATION TECHNIQUES

Statman and Glushkov (2016) find that the two main portfolio construction methods (ESG score screening and industry filtering) do not deliver the same results. They find value for the former but not for the latter. Mohanty et al. (2021) introduce ESG target. They construct overlays by over weighting stocks with higher ESG metrics until the portfolio improves its ESG score by 20% compared to a given benchmark. Gurvich and Creamer (2021) show that several carbon emission based sorts lead to various outcomes. Depending on whether portfolios are built on raw emissions (which introduces a size bias), emissions divided by market capitalization, or emissions divided by sales, the Sharpe ratios shift from good to outstanding.

Branch and Cai (2012) propose an original idea: build ESG portfolios that mimic the S&P500. They show it is possible to deliver market performance with portfolios relying only on socially responsible stocks. Fan and Michalski (2020) improve momentum and quality factors with ESG screens and sorting procedures. Henriksson et al. (2019) introduce an original solution to overcome issues when firms do not have ESG ratings. They advocate the creation of an ESG factor (Good minus Bad) and the evaluation of each stock's exposure to this factor. A loading significantly above zero should mean that the firm's returns are positively driven by the ESG factor. At the aggregate level, the aim is to tilt portfolios toward *good* ESG proxies. Khan (2019) proposes a new methodology to construct robust governance and ESG scores that seem to do a good job at explaining returns. Palazzolo et al. (2020) discuss the challenges in the crafting of carbon neutral portfolios.

In Widyawati (2020) (p. 633), the author makes the case that *"there is little empirical knowledge about the most effective way to apply SRI as a quantitative financial model and how this application affects market equilibrium."* It seems more reasonable to conclude that there is a large body of empirical work, but the problem is that conclusions of some studies often contradict the findings of other analyses. There is no consensus on how precisely to efficiently integrate ESG ingredients in a portfolio strategy that performs well financially out-of-sample.[2] There is probably less uncertainty in the social good that comes from SRI, compared to its pecuniary benefits, although Cappucci (2018) warns against the perils of *half measures* that are only implemented for marketing purposes (see also Statman (2020) on this matter).

At an aggregate level, Harper (2020) provides some tips on how to integrate ESG criteria when searching and selecting an investment manager. Schoenmaker and Schramade (2019) draw the contours of an alternative investment paradigm for investors who seek long-term value creation. Cosemans et al. (2021) specify a VAR model that includes temperature change as a predictor for (aggregate) equity returns. They estimate their model with a Bayesian method and subsequently maximize a constant relative risk aversion (CRRA) utility function based on these beliefs. Umar et al. (2020) document that ESG indices are interconnected at the global level, thereby challenging the benefits of geographical diversification. Finally, Parker (2021) wraps the notion of impact investing in a goal-based

---

[2]In fact, delivering relative outperformance is hard, even without any ESG constraints at all.

allocation framework to help balance financial and sustainability targets.

One very interesting contribution is the work of Raynaud et al. (2020). The goal of the article is to articulate a methodology for portfolio managers who want to craft their allocation so as to participate to the global initiative toward the 2°C alignment (fixed by the 2015 Paris accord). The paper links climate metrics, macro-economics scenarios, and portfolio engineering in a non-technical and insightful fashion. For more details, the interested reader can have a peek at the *alignment cookbook* written by the authors. The assessment of the carbon footprint of portfolios, a material topic, is discussed in Erlandsson (2021).

Finally, the first issues of the *Journal of Impact and ESG Investing* are filled with various pieces of advice. For instance, Grim and Berkowitz (2020) provide general guidance when including ESG criteria in the investment process and mention a few use cases. Chan et al. (2020) show how to use intangible value and corporate culture proxies to improve the performance of more traditional factors (value and quality).

CHAPTER 6

Climate change risk

"Climate risk is investment risk" (Scanlan (2021)).

This chapter is dedicated to the perils of global warming and their impact on firms and on the financial market in general.[1] The importance of this threat has long been mostly overlooked but is now even documented by governmental commissions in the US (see Behnam et al. (2020)). In fact, as early as the mid-1990s, Porter and Van der Linde (1995) called for a change of paradigm in the trade-off between environmental friendliness and competitiveness. At the time, the authors based their arguments on the need to incorporate innovation (and its dynamics) as a key variable. Nowadays, this trade-off seems more focused on long-term risks related to climate change (see, e.g., Daniel et al. (2018), Barnett et al. (2020)).

We start this chapter by simply quoting the provocative paper of Mayer (2019): *"natural capital is very different from other forms of capital and arguably should not be viewed as a capital at all. Its distinctive features are its renewable and restorative properties, its irreversibility, its living and evolving nature, and the fact that it was inherited, not created, by humans."* Climate risks are now documented and for instance surveyed in Breitenstein et al. (2021) and Giglio et al. (2021). Nevertheless, in their survey of experts, Stroebel and Wurgler

---

[1]Several theoretical references are postponed to the online Chapter.

(2021) find that the latter overwhelmingly believe that asset prices underestimate climate risks. According to them, the short term risk is **regulatory** (e.g., benefits being curtailed by carbon taxes), while the long term risk is **physical** (e.g., natural disasters impacting production, transport, or consumption of goods). The regulatory risk, combined to risks related to shifts in consumer preferences for instance are aggregated into what are often referred to as **transition risks**.

Luckily, international cooperation on the matter seems to be intensifying (Carattini et al. (2021))!

This chapter is divided in four separate parts. The first part deals with discounting utility and cash flows in uncertain environments. The second part covers measurement issues in the assessment of climate change. The third part gives a quick overview of some macro-economic impacts of global warming. Finally, the last subsection demonstrates that investors increasingly care about these issues (which echoes Section 3.1).

6.1 UNCERTAIN DISCOUNTING

The mathematics-averse reader is advised to skip this subsection.

Discounting is a central topic in financial analysis because it translates the value of future flows into current units. Depending on preference and beliefs, discounting factors will alter expected cash flows and utility. In this subsection, we briefly recall how uncertainty may affect returns. We start by recalling the model of Ramsey (1928).

An entity (individual or society) seeks to optimize a definition of global welfare

$$W = \sum_{t=0}^{\infty} \beta^t u(c_t), \qquad (6.1)$$

where $u(\cdot)$ is some utility function, c_t is time-t consumption, and $\beta \in (0,1)$ is the discounting intensity. A low beta signals a strong preference for the most imminent dates, while a high beta puts a higher weight on the distant future. Sometimes, the conventions $\beta = (1+\rho)^{-1}$ and $\beta = e^{-\delta}$ are used, in which case ρ and δ are the discount rates. The entity

1. has capital wealth k_t which it can invest on a financial asset,
2. earns a wage $w_t > 0$, and
3. faces a budget constraint:[2]

$$k_{t+1} = e^{r_t} k_t + w_t - c_t \geq 0, \qquad (6.2)$$

that is, future wealth equals wealth invested on the asset at (log-)rate r_t, plus wage, minus consumption. It is naturally assumed that the wealth remains positive.

The entity must choose the levels of investment k_t and consumption c_t to maximize the welfare W. Taking into consideration two consecutive points in time, the (restricted) Lagrangian reads

$$\mathcal{L} = \beta^t u(c_t) - \lambda_t (k_{t+1} - e^{r_t} k_t - w_t + c_t) \qquad \text{(time } t)$$
$$(6.3)$$
$$+ \beta^{t+1} u(c_{t+1}) - \lambda_{t+1}(k_{t+2} - e^{r_{t+1}} k_{t+1} - w_{t+1} + c_{t+1}), \quad \text{(time } t+1)$$

and the first-order conditions command

$$\frac{\partial \mathcal{L}}{\partial c_t} = \beta^t u'(c_t) - \lambda_t = 0, \quad \text{and} \quad \frac{\partial \mathcal{L}}{\partial k_{t+1}} = -\lambda_t + \lambda_{t+1} e^{r_{t+1}},$$
$$(6.4)$$

and plugging the left part into the right one, this translates to

$$\beta^t u'(c_t) = \beta^{t+1} u'(c_{t+1}) e^{r_{t+1}} \iff \frac{u'(c_t)}{u'(c_{t+1})} = \beta e^{r_{t+1}}. \quad (6.5)$$

This equation is a cornerstone of consumption-based asset pricing (see Cochrane (2009)) because it links returns to the ratio of marginal utilities (present versus future). While the

---

[2]Depending on how r_t is defined (logarithmic versus arithmetic return), the following constraint can also be found in the litterature $k_{t+1} = (1 + r_t)k_t + w_t - c_t \geq 0$. In Equation (6.2), r_t is the log return of the investment opportunity (on the financial market).

asset pricing literature usually takes the route of the pricing kernel (or stochastic discount factor), we pursue the analysis as it is derived in economics.

Often, the utility function is chosen to be CRRA, so that $u(x) = x^{1-\alpha}/(1-\alpha)$, for α strictly positive but not equal to one (in the latter case, the logarithmic function is used instead). In this case, with $\beta = e^{-\delta}$, and

$$\left(\frac{c_t}{c_{t+1}}\right)^{-\alpha} = e^{r_{t+1}-\delta} \iff r_{t+1} = \delta + \alpha \log\left(\frac{c_{t+1}}{c_t}\right). \quad (6.6)$$

The above rate r_{t+1} is sometimes referred to as the *social discounting rate* (SDR) in the public economics literature. It is such that the entity is indifferent between the two options:

1. consume more at time t and enjoy immediate utility or
2. reduce consumption and invest to gain more at time $t+1$, with a discount of δ.

Another way to interpret the result is that the return on investment on the left-hand side in Equation (6.6) must be equal to the welfare-preserving inter-temporal trade-off on the consumption (right-hand side). If consumption growth increases, then the SDR should also increase, in order to cover the future consumption needs. The SDR is very important because it is a crucial component in the computation of the social cost of carbon (see Anthoff et al. (2009) and the online Chapter). It is used to discount (in time) the welfare of a population, often by attenuating the importance of aggregate utility (based on consumption) as time passes.

One interesting extension of this model pertains to the alteration of this social discount rate when consumption growth is random.[3] This idea has gained traction at least since Gollier (2002), and they are applied to a climate change paradigm in Gollier (2013).[4] Let us assume that $g_{t+1} = \log(c_{t+1}/c_t)$ is

---

[3]According to Drupp et al. (2018), experts seem to agree that the simple Ramsey Equation (6.6) is too limited.

[4]Further theoretical results in this vein can be found in Gollier (2008),

Gaussian with mean μ and variance σ^2. Under CRRA preferences, at time t, we can take the expectation of Equation (6.5) as follows:

$$e^{\delta - r_{t+1}} = \mathbb{E}[e^{-\alpha g_{t+1}}] \implies r_{t+1} = \delta + \alpha \mu - \alpha^2 \frac{\sigma^2}{2}, \qquad (6.7)$$

which means that uncertainty in future consumption decreases the return. Straightforwardly, any investor prefers *less* risk. For the interested reader, we recommend a few additional references on the topic of climate economics: Ackerman et al. (2009), Heal (2009), Weitzman (2009), and, more recently, Gollier (2021) on the topic of carbon prices. For a recent discussion on estimation issues, we point to Newell et al. (2021).

Lastly, on a related issue, Gelrud (2021) produces a theoretical model aimed at quantifying the rate at which climate change mitigation project should be discounted. The authors shows that this rate should be smaller than the risk free rate - and that it is optimal to invest in such projects as fast as possible.

6.2 MEASUREMENT ISSUES

In order to quantify the risk of global warming, it is first imperative to define which variables drive the externalities. They can be direct measurements of the underlying phenomenon (e.g., local or aggregate temperatures and rainfall), or time-series of potential drivers thereof (greenhouse gas (GHG) and carbon dioxide (CO_2) emissions). In addition to measuring, it is also useful to predict or even nowcast such quantities: see Bennedsen et al. (2021) for a methodology on CO_2 emissions. Prediction models are important because they seem to drive market participants' expectations (Schlenker and Taylor (2021)). International groups, such as the Intergovernmental Panel on Climate Change (IPCC) periodically disclose

---

Traeger (2014) and Fleurbaey and Zuber (2015). Applications to the impact of climate change on the economy have spawned a rich body of articles. Here are a few: Howarth (2003), Dasgupta (2008), and the review by Heal (2009).

in-depth studies that contain numerous estimates on past, present and future indicators (emissions, temperatures, precipitations). On the topic of climate data, we also refer to Tankov and Tantet (2019) for a very enlightening discussion on the dimensions and stakes for financial agents.

Measuring and reporting climate-related indicators requires resources, which is why it is often performed by national or international research centers. Per se, the equipment (thermometers, CO_2 sensors, etc.) are not particularly expensive. It is keeping track of trustworthy measurements over long time ranges which is costly. With regard to the evolution of CO_2, one benchmark is the measurement by the US Earth System Research Laboratories near the summit of the Mauna Loa volcano.[5] The corresponding time-series is shown in the top panel of Figure 6.1 and shows an indisputable trend.

Recent initiatives propose measures at a more granular (i.e., local) level, see for instance Liu et al. (2020) and Dou et al. (2021) as well as the website https://carbonmonitor.org, as well as climateestimate.net. The latter proposes code snippets in R, Python and MATLAB. This allows to track emissions at the country and sector level and we provide a sample of trajectories for 2019-2021 in Figure 6.2. Clearly, the reduction in emissions appears in China (which was hit earlier) a few weeks before Europe and the US for aviation and ground transport. For industry and power, the impact is more pronounced for China, but the rebound is also marked.

At the firm level, we refer to https://climatechangelab.info. The evaluation of climate change exposure is postponed to section 6.4.

For temperatures, the challenge is different because of the dimension of the reporting. Thousands of thermometers track variations in meteorological stations and it is cumbersome to access and aggregate such data. Most sensors are located on land, which only covers 29% of the globe's surfaces. Thus

---

[5]Technical details on the measurement can be found in Zhao and Tans (2006).

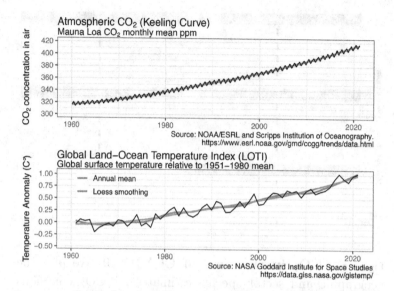

Figure 6.1 **Sample of climate related time-series.** We plot atmospheric CO_2 levels along with global temperature trends over the period 1960–2021. The data was gathered thanks to the *hockeystick* **R** package.

ocean temperature must also be measured (at their surfaces), and this is performed with ships. We refer to the paper by Hansen et al. (2010) for a precise account on this matter. In the lower panel of Figure 6.1, we plot the evolution of the *global* temperature of our planet. While the curve is less regularly increasing, the trend is again undeniable. The parallel display of the two series reveals a strong *correlation*. It is now widely accepted that the relationship is *causal*, whereby CO_2 emissions are a key driver of the increase in temperature. We point to Le Treut (2007) for a historical perspective on this debate. More recent contributions document feedback effects (Van Nes et al. (2015)), or clear direct causality (Stips et al. (2016)).

From an investment standpoint, being able to quantify physical risks linked to climate change is important. Hain et al. (2021) show that, just like for traditional ESG metrics,

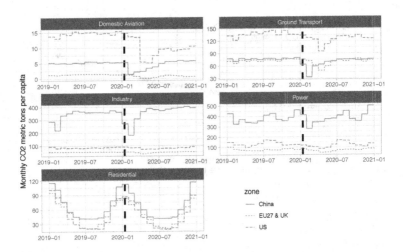

Figure 6.2 **CO_2 in the time of COVID-19**. We plot the geographic and sector specific estimates for CO_2 emissions provided by https://carbonmonitor.org. The vertical dashed line marks January 13th, 2020, which is the date when a first case of COVID-19 was discovered outside China (in Thailand).

ratings of physical risks are very much provider-dependent. They compare series from Trucost, Carbone 4 Finance, Southpole, Truevalue Labs, and two academic-based measures and find that correlations within sectors are relatively small. This implies that portfolios sorted on these metrics have limited overlaps when switching from one provider to another.

6.3 STRESS TESTS AND OTHER MEASURES

In Battiston et al. (2017), GHG emission data is used to stress-test the financial system. In Reinders et al. (2020), the focus of the stress test study is set on taxes and their impacts. The risk for firms is that governments will be increasingly inclined to penalize polluters, thereby threatening parts of their balance sheets. In their proposal for an integrated stress-testing methodology, Allen et al. (2020) promote a scenario-driven approach in which climate models are intertwined with macro-economic models. The procedure is able to generate individual firms' probabilities of defaults, as well as market valuations. A similar framework is adopted by Fang et al. (2019), and subsequently combined to mean-variance optimization to build portfolios that are less sensitive to climate change risk. Relatedly, Monasterolo (2020) proposes new climate risk metrics, such as the climate Value-at-Risk, which is computed on corporations that would be most affected in stress scenarios.

In their attempt to measure climate change risk for corporations, Li et al. (2020) proceed very differently: they extract textual sentiment from earning call transcripts. The authors use a dedicated climate-centric lexicon combined to manual verifications. From this, they synthesize several climate risk measures which they study in detail. Chou and Kimbrough (2019) automate the textual screening of firms' SEC filings. They show that the frequency at which climate change terms are mentioned increases through time, though the patterns differ from one industry to the other. Faccini et al. (2021) also analyze textual factors and split risks in two categories: **transition risk** (via policy (e.g., fiscal) or shifts in consumer preferences) and **physical risk** (e.g., catastrophes). They find that only the first one is priced in the US stock market.

Hsu and Wang (2013) also resort to sentiment analysis, but focused on one media outlet (*The Wall Street Journal*). They measure the negativity of the tone related to climate change articles in the press. Surprisingly, they report that the aggregate market reacts positively to negative news. Natural language processing (NLP) is also exploited in Engle et al. (2020) to build portfolios that hedge investors against negative news related to climate change. In a similar vein, Sautner et al. (2021a) use transcripts of earnings conference calls to assess the exposure of 10,000 firms to opportunity, physical, and regulatory shocks associated with global warming. They use their methodology in a follow-up paper that links climate risk exposure to risk premia. In a different setting, Heo (2021) also uses their data to reveal that firms that are more exposed to climate change risk increase their cash holdings, probably in anticipation of adverse situations. Relatedly, Santi (2021) crafts a climate sentiment index and shows how it dynamically affects the performance of an Emission-minus-Clean portfolio.

For a more computer science-focused perspective on NLP-driven classification of ESG topics and climate risks, we refer to Nugent et al. (2020), Raman et al. (2020), Amel-Zadeh et al. (2021) (to measure corporate alignment with SDGs), Bingler et al. (2021) (on the so-called *ClimateBERT*), Apel et al. (2021), Borms et al. (2021), and Sokolov et al. (2021).

Lastly, the *Journal of Financial Stability* dedicated a special issue on the role of climate risk on financial perturbations (see Battiston et al. (2021)).

6.4 MICRO- AND MACRO-ECONOMIC IMPACTS

Climate events or disasters are susceptible to affect firms in numerous ways. In Addoum et al. (2019) and Hugon and Law (2019)), the abnormal heat is linked to reductions in earnings, while in Gostlow (2019) it is shown that a rainfall factor explains the cross-section of stocks in the US, Europe, and Japan. In Liu and Xu (2017), the air quality index is shown to have strong effects on the stock markets, both at the individual firm level and at the aggregate level. Tol (2021) disentangles the impacts of climate (long-term changes) and weather (temporary shocks) on the economy (productivity, i.e., production output per worker is the dependent variable). The paper shows that both matter, especially long-run temperatures (climate) and abnormal precipitation (weather). Rising sea levels are also a major concern (see Bernstein et al. (2019)). More and more, indices are being developed to capture or predict the effects of global warming. For instance, Jiang and Weng (2020) rely on the Actuaries Climate Index to build efficient portfolios of agriculture-related firms. Lastly, in a related investment field, the impact of climate change on the real estate market is abundantly documented, but is out of the scope of this review, though it does have direct consequences for financial markets.

Naturally, for investors, being able to gauge if a firm is exposed to climate change has become crucial. There is of course no unique way to proceed. For instance, Sautner et al. (2021a) resort to a machine learning analysis of corporate conference calls. Görgen et al. (2020) and Roncalli et al. (2020) measure exposure via regressions against brown-minus-green (BMG) risk factor.

We now split the contributions on macro-economic impacts in three categories: event studies, disaster modelling and statistical analysis. In the former, special and punctual events are scrutinized and researchers try to evaluate if changes have occurred after the event, e.g., if trends have stopped or reversed. More generally, valuable reference on this topic is the book by Tol (2019).

For instance, according to Monasterolo and De Angelis (2020), investors have had more consideration for low carbon assets after the Paris Agreement. Ramelli et al. (2020) report that the success of the Global Climate Strike in March 2019 has increased expectations of investors toward carbon intensive firms (i.e., it has pushed their cost of capital upward). Sen and von Schickfus (2020) studies the progressive impact of German environmental policies on utility companies. They find early policies had no effect but later ones did, pointing to a risk of stranded assets perceived by investors. Relatedly, Ma et al. (2021) find that stocks co-move with the market more during climate disasters.

In disaster models (see also the online Chapter), authors investigate how externalities can affect firm cash flows, risk, or investor demands. Mittnik et al. (2020) for instance document the impact of climate-related shocks on capital losses. Relatedly, Lanfear et al. (2019, 2020) find that stocks underperform during hurricane events, with the exception of high tech companies.

Once the risk is acknowledged, investors should take it into account. Shen et al. (2019) propose an asset allocation scheme based on a large VAR(1) estimation in which changes in temperatures are used as state variables. Kumar et al. (2019) also exploit climate-related information to build profitable long-short portfolios. They show that stocks that are more sensitive to abnormal temperature changes earn lower returns than those that are less sensitive to these temperature variations.

Finally, Kahn et al. (2019) use a large-scale panel data analysis to link the per-capita real output growth to changes in temperatures. They find that exposures are strongly country-dependent. Colacito et al. (2019) find that a 1°F increase in summer temperature reduces state-output growth in the US by 0.15 to 0.25 percentage points. Using quantile regressions, Kiley (2021) document a significant link between temperatures and economic downside risks (strong contractions of the GDP per capita).

6.5 INVESTOR ATTENTION

In their large scale survey of investor preferences, Ilhan et al. (2021) document several salient trends. First, they find that a majority of investors are willing to disclose the carbon impact of their portfolios. In fact, many consider climate risk reporting to be as important as financial reporting. This translates into actions because higher institutional ownership in firms is linked to higher propensity to voluntarily disclose carbon emissions and to provide higher quality information. In the same vein, Anderson and Robinson (2020) report a shift in individual investor beliefs after extreme weather events in Sweden in 2014. Posterior to the climate calamities, these investors shifted their retirement portfolios toward sustainable funds. Similarly, Makridis (2021) extreme temperatures distort investors' beliefs on aggregate growth, thereby altering asset prices. It is found that days with abnormally hot or cold temperatures experience lower stock returns. Using data from the Spatial Hazards Events and Losses Database for the United States (SHELDUS), Marshall et al. (2021) also find that, after climate disasters, investors shift resources towards more environmentally-friendly mutual funds.

At a more macro-level, Wu and Lu (2020) build a search engine-based index that captures the mood of investors, and they detail how it impacts the market liquidity and volatility. Choi et al. (2020) also resort to search engine data (the Google Search Volume Index) to evaluate if people's attention is shifted by shocks to local temperatures. When the weather is unusually hot, carbon-intensive firms experience lower returns, compared to greener firms. Bessec and Fouquau (2020) scan article from the *Wall Street Journal* to build sentiment indices with respect to environment issues. They show that sectors are impacted differently by variations in these indices. In Alok et al. (2020), investor perception is examined through the prism of natural disasters. The authors find that funds located close to disaster areas reduce their portfolio holdings in firms located close to this area. Finally, Fiordelisi et al. (2021) find that ESG ETFs perform well after episodes of climate disasters and conclude that this must be because investors reallocate towards SRI funds subsequently to periods of climate tensions.

6.6 POLICY

Naturally, given the stakes induced by climate change, economists have contributed to the debate by proposing policies to reduce the impact of global warming. Themes, scopes and methods are diverse; we provide a very brief overview below. With respect to abatement policies, the main reference is the survey of Pindyck (2013).

A major topic when thinking about climate change mitigation is the role of regulators. The most straightforward policy measure is the carbon tax, whereby firms would have to pay a fee depending on their level of carbon emissions. Several books are dedicated to this subject (Hsu (2012), Milne and Andersen (2012), Kreiser et al. (2015), Cramton et al. (2017), Metcalf (2018)), but we recommend the public handbook by the World Bank (Partnership for Market Readiness (2017)).

The most decisive question on this matter is: what is the impact of carbon taxes on the economy? Unfortunately, carbon taxes remain marginal and they are enforced on rather small scales, which means that there is not an abundance of data to help researchers answer the question. Below, we list a few attempts in this direction.

- **No impact on firms**. According to Venmans et al. (2020), carbon taxes have not been shown to be antithetical to competitiveness.
- **Positive effect on the economy**. The article Porter and Van der Linde (1995) is an early contribution that proposes that environmental regulations may be good for firms and competitiveness because it fosters innovation (this is often referred to as the *Porter hypothesis*). It has for instance been confirmed in Quebec (Lanoie et al. (2008)), in the OECD (Lanoie et al. (2011)), and in Europe (Costantini and Mazzanti (2012)). Brown et al. (2021) show that, at least, carbon taxes are a strong incentive for polluting firms to spend more on R&D - though it's not clear that this effort is environmentally focused.
- **It depends**. The early survey of Bosquet (2000) establishes that the impact depends on the time horizons. Benefits seem to occur in the short term, but effects in the long term are uncertain. Klenert et al. (2018) come to the conclusion that the optimal modality of carbon pricing depends on the political context (e.g., on the level of trust in the government or on the main concerns of the population). King et al. (2019) dissect the implications of carbon taxes at the sector level. They show that targeted taxes are the most efficient. In their overview of emission trading systems, Narassimhan et al. (2018) reveal that the success of carbon pricing initiatives can depend on several factors, such as administrative prudence, appropriate carbon revenue management and stakeholder engagement.

- **Positive effect on the environment**. Rafaty et al. (2021) contend that carbon taxes have reduced the growth of CO_2 emissions by 1 to 2.5% and the authors conclude: "*carbon pricing alone is unlikely to be sufficient to achieve emission reductions consistent with the Paris climate agreement*." This conclusion is also supported by the meta-analyses Patuelli et al. (2005) and Green (2021).

- **Negative effect on the economy**. In their study on California's cap-and-trade program, Bartram et al. (2021) report that climate policies can backfire. They show that financially unconstrained firms do not reduce their total emissions. What is worse is that they find that constrained firms increase total emissions by shifting activity to states that are not subject to carbon penalties.

In addition, several theoretical models tackle the topic of carbon pricing. For instance, Finkelstein Shapiro and Metcalf (2021) propose an equilibrium model in which the introduction of a carbon tax has *positive* effects on labor income, consumption, output and labor force participation, but has a negative impact on employment. A different angle is studied in Aïd and Biagini (2021), who also discuss carbon emission reduction, but in this paper, regulatory instances pursue their objectives by allowing emission permits. Both firms and the regulator optimize their own utility function to reach an equilibrium. One notable finding is that optimal abatement efforts are constant through time. Furthermore, Traeger (2021) studies the impact of uncertainty (emissions, temperature, and damages) on the optimal level of the carbon tax (considered to be the social cost of carbon (SCC), see also the online Chapter). The paper shows that in the absence of uncertainty, the tax is linear in several factors (e.g., climate variables), but that the relationship becomes convex when the climate state variables are stochastic. Benmir et al. (2020) conclude that the optimal tax level depends on the shadow price of emissions and that it it pro-cyclical. It should be high during booms to cool the economy down, and low during recessions to stimulate a recovery.

Dunz et al. (2021) build a complex economic model in which banks are sensitive to a green sentiment. This specification allows banks to possibly anticipate a carbon tax or a green supporting factor and increase its loan rates to brown companies. The authors show a carbon tax would be more efficient if it is combined with redistributive and welfare measures and if banks indeed resort to a climate sentiment so that the transition to a low carbon economy is smooth. Finally, Lemoine and Traeger (2014) analyze the optimal level of carbon taxes in a world with tipping points, which are moments of time when damage to the planet is irreversible and may cause chain reactions. According to their projections, the intensity of carbon taxes should increase until 2150, and then possibly decrease.

Beyond carbon pricing, there are other ways to curb the economic activity toward more sustainability. For example, Yao et al. (2021) analyze the impact of green credit policies (when banks favor green projects and firms when allowing loans) and find (unsurprisingly) that it is only penalizing for heavily polluting industries (e.g., electrolytic aluminum, petrochemical and tanning). In their study on Chinese markets, Zhang et al. (2021) find that ESG firms experience higher returns (compared to low ESG stocks) since 2016, when green policies were enforced. This shows that governmental action can be key to promote certain types of assets. McKibbin et al. (2020) argue that climate change and the resulting shocks are likely to alter central banks' ability to predict and manage inflation. The authors mention challenges in the conception of joint climate and monetary policies.

The angle of monetary policy is mentioned in several studies. Dafermos et al. (2018) come to pessimistic conclusions about the risk that climate change induces on economies and financial markets, in particular. Nevertheless, they contend that green quantitative easing can help reduce financial instability. Dietrich et al. (2021) argue that natural disasters reduce the natural rate of interest. After analyzing a survey on expectations toward climate change and building a New Keynesian model, they also find that climate risk shrinks inflation and the output gap by 0.3 and 0.2%, respectively.

CHAPTER 7

Conclusion

At the heart of ESG investing lies a dichotomy between sustainability and profitability. Many researchers claim that it is possible to obtain both at the same time, popularizing the *"doing good and doing well"* motto. When we span hundreds of articles, the conclusions are not clear-cut, which is the conclusion of most meta-studies. But at the same time, if SRI is (on average) not costly and may be socially beneficial, then it almost qualifies as ethical free lunch. This is the most salient takeaway from the book. Over the past two decades, there is no strong indisputable evidence that sustainable investing is financially detrimental, which is why investors who care (even remotely) about the long-term future should not hesitate one second to embrace SRI.

At the beginning of the 2020 decade, green stocks have performed relatively well. They may become expensive, which would lower their expected returns. Investors should not be disappointed if realized returns of green indices become less attractive, compared to conventional indices. The green premium (if there is one) will fluctuate.

There are, however, signs that investors are shifting resources toward green firms simply because brown firms are increasingly perceived as bearing risks that are both hard to evaluate and potentially devastating. This is meaningful in the aftermaths of major crises because investors want to attenuate the systemic risks in their portfolios. In the long run, the multiplication of natural disasters is likely to force governments to fiscally penalize polluters. Companies with increased exposure to physical risks (heat waves, rising sea levels) and transition risks (carbon taxes) will see their cost of capital soar. The rise of green activism and lobbying is also a threat for brown corporations. More and more, sustainable firms may be considered as safer bets, rightfully so.

Nevertheless, it is not obvious that the rise of ESG investing has a significant impact globally. First, by leaving brown firms out of the loop, green investors miss an opportunity to bend the policies of the worst-in-class corporations toward increased sustainability. In fact, polluting firms can sometimes be the ones that invest the most in sustainable technologies. Moreover, it is premature to conclude whether individuals or institutions who invest in green firms participate actively in the pursuit of ESG-compatible goals. This topic is a fertile ground for scholars who carry out research in shareholder activism.

Future research will need to identify more precisely which determinants matter most from a financial and social standpoint. Granular characteristics (on emissions, governance and social issues) are increasingly being disclosed by firms, and they open the way for finer analyses. Uniform regulations enforcing strict reporting constraints would benefit all stakeholders by imposing transparency in a discipline that needs it desperately.

Bibliography

Abate, G., I. Basile, and P. Ferrari (2021). The level of sustainability and mutual fund performance in Europe: An empirical analysis using ESG ratings. *Corporate Social Responsibility and Environmental Management 28*(5), 1446–1455.

Abdi, Y., X. Li, and X. Càmara-Turull (2021). Exploring the impact of sustainability (ESG) disclosure on firm value and financial performance (FP) in airline industry: the moderating role of size and age. *Environment, Development and Sustainability*, 1–28.

Abeysekera, A. P. and C. S. Fernando (2020). Corporate social responsibility versus corporate shareholder responsibility: A family firm perspective. *Journal of Corporate Finance 61*, 101370.

Abhayawansa, S. and S. Tyagi (2021). Sustainable investing: The black box of environmental, social and governance (ESG) ratings. *Journal of Wealth Management 24*(1), 49–54.

Ackerman, F., S. J. DeCanio, R. B. Howarth, and K. Sheeran (2009). Limitations of integrated assessment models of climate change. *Climatic change 95*(3-4), 297–315.

Addoum, J. M., D. T. Ng, and A. Ortiz-Bobea (2019). Temperature shocks and industry earnings news. *SSRN Working Paper 3480695*.

Adedeji, B. S., O. M. J. Popoola, and S. O. Tze (2017). National culture and sustainability disclosure practices: A literature review. *Indian-Pacific Journal of Accounting and Finance 1*(1), 26–50.

Adriaan Boermans, M. and R. Galema (2020). Carbon home bias of European investors. *SSRN Working Paper 3632723.*

Agrawal, A. and K. Hockerts (2019a). Impact investing: review and research agenda. *Journal of Small Business & Entrepreneurship*, 1–29.

Agrawal, A. and K. Hockerts (2019b). Impact investing strategy: managing conflicts between impact investor and investee social enterprise. *Sustainability 11*(15), 4117.

Ahsan, T., B. Al-Gamrh, and S. S. Mirza (2021). Corporate social responsibility and firm-value: the role of sensitive industries and CEOs power in China. *Applied Economics*, 1–20.

Aïd, R. and S. Biagini (2021). Optimal dynamic regulation of carbon emissions market: A variational approach. *arXiv Preprint* (2102.12423).

Akhtar, F., M. Veeraraghavan, and L. Zolotoy (2021). Board gender diversity and firm value in times of crisis: Evidence from the COVID-19 pandemic. *SSRN Working Paper 3869585.*

Alareeni, B. A. and A. Hamdan (2020). ESG impact on performance of US S&P 500-listed firms. *Corporate Governance: The International Journal of Business in Society 20*(7), 1409–1428.

Alda, M. (2021). The Environmental, Social, and Governance (ESG) dimension of firms in which Social Responsible Investment (SRI) and conventional pension funds invest: the mainstream SRI and the ESG inclusion. *Journal of Cleaner Production*, 126812.

Alessandrini, F. and E. Jondeau (2020). ESG investing: From sin stocks to smart beta. *Journal of Portfolio Management 46*(3), 75–94.

Alessandrini, F. and E. Jondeau (2021). Optimal strategies for ESG portfolios. *Journal of Portfolio Management 47*(6), 114–138.

Alessi, L., E. Ossola, and R. Panzica (2020). The Greenium matters: evidence on the pricing of climate risk. *SSRN Working Paper 3452649*.

Alessi, L., E. Ossola, and R. Panzica (2021). What greenium matters in the stock market? The role of greenhouse gas emissions and environmental disclosures. *Journal of Financial Stability 54*, 100869.

Alexandridis, G., A. G. Hoepner, Z. Huang, and I. Oikonomou (2021). Corporate social responsibility culture and international M&As. *British Accounting Review Forthcoming*, 101035.

Alfonso-Ercan, C. (2020). Private Equity and ESG investing. In *Values at Work*, pp. 127–141. Springer.

Allen, T., S. Dees, V. Chouard, L. Clerc, A. de Gaye, A. Devulder, S. Diot, N. Lisack, F. Pegoraro, M. Rabate, et al. (2020). Climate-related scenarios for financial stability assessment: an application to france. *SSRN Working Paper 3653131*.

Alok, S., N. Kumar, and R. Wermers (2020). Do fund managers misestimate climatic disaster risk. *Review of Financial Studies 33*(3), 1146–1183.

Alshater, M. M., O. F. Atayah, and A. Hamdan (2021). Journal of sustainable finance and investment: A bibliometric analysis. *Journal of Sustainable Finance & Investment*, 1–22.

Ambec, S. and P. Lanoie (2008). Does it pay to be green? A systematic overview. *Academy of Management Perspectives*, 45–62.

Amel-Zadeh, A. (2021). The materiality of climate risk. *SSRN Working Paper 3295184*.

Amel-Zadeh, A., M. Chen, G. Mussalli, and M. Weinberg (2021). NLP for SDGs: Measuring corporate alignment with the sustainable development goals. *SSRN Working Paper 3874442*.

Amel-Zadeh, A. and G. Serafeim (2018). Why and how investors use ESG information: Evidence from a global survey. *Financial Analysts Journal 74*(3), 87–103.

Amihud, Y., M. Schmid, and S. D. Solomon (2017). Settling the staggered board debate. *U. Pa. L. Rev. 166*, 1475.

Ammann, M., D. Oesch, and M. M. Schmid (2011). Corporate governance and firm value: International evidence. *Journal of Empirical Finance 18*(1), 36–55.

Amon, J., M. Rammerstorfer, and K. Weinmayer (2021). Passive ESG portfolio management - The benchmark strategy for socially responsible investors. *Sustainability 13*(16), 9388.

Anderson, A. and D. T. Robinson (2020). Talking about the weather: Availabilty, affect and the demand for green investments. *SSRN Working Paper 3490730*.

Anson, M., D. Spalding, K. Kwait, and J. Delano (2020). The sustainability conundrum. *Journal of Portfolio Management 46*(4), 124–138.

Anthoff, D., R. S. Tol, and G. W. Yohe (2009). Discounting for climate change. *Economics 3*, 1–22.

Antoncic, M., G. Bekaert, R. V. Rothenberg, and M. Noguer (2020). Sustainable investment - Exploring the linkage between alpha, ESG, and SDG's. *SSRN Working Paper 3623459.*

Apel, M., A. Betzer, and B. Scherer (2021). Real-time transition risk. *SSRN Working Paper 3911346.*

Aragon, G. O., Y. Jiang, J. Joenväärä, and C. I. Tiu (2020). Socially responsible investments: Costs and benefits for university endowment funds. *SSRN Working Paper 3446252.*

Ardia, D., K. Bluteau, K. Boudt, and K. Inghelbrecht (2020). Climate change concerns and the performance of green versus brown stocks. *SSRN Working Paper 3717722.*

Arjaliès, D.-L. (2010). A social movement perspective on finance: How socially responsible investment mattered. *Journal of Business Ethics 92*(1), 57–78.

Arjaliès, D.-L. and P. Bansal (2018). Beyond numbers: How investment managers accommodate societal issues in financial decisions. *Organization Studies 39*(5-6), 691–719.

Arlow, P. and M. J. Gannon (1982). Social responsiveness, corporate structure, and economic performance. *Academy of Management Review 7*(2), 235–241.

Arnold, M. C., C. Hörner, P. R. Martin, and D. V. Moser (2020). German and US investment professionals' use of corporate social responsibility disclosures in their personal investment decisions and recommendations to clients. *SSRN Working Paper 3020887.*

Arora, S., J. K. Sur, and Y. Chauhan (2021). Does corporate social responsibility affect shareholder value? Evidence from the COVID-19 crisis. *International Review of Finance Forthcoming.*

Arouri, M., M. Gomes, and K. Pukthuanthong (2019). Corporate social responsibility and m&a uncertainty. *Journal of Corporate Finance 56*, 176–198.

Arribas, I., M. D. Espinós-Vañó, F. García, and N. Riley (2021). Do irresponsible corporate activities prevent membership in sustainable stock indices? the case of the Dow Jones Sustainability Index World. *Journal of Cleaner Production 298*, 126711.

Artiach, T., D. Lee, D. Nelson, and J. Walker (2010). The determinants of corporate sustainability performance. *Accounting & Finance 50*(1), 31–51.

Ashbaugh-Skaife, H., D. W. Collins, and R. LaFond (2006). The effects of corporate governance on firms' credit ratings. *Journal of Accounting and Economics 42*(1-2), 203–243.

Aslan, A., L. Poppe, and P. Posch (2021). Are sustainable companies more likely to default? Evidence from the dynamics between credit and ESG ratings. *Sustainability 13*(15), 8568.

Aswani, J., A. Raghunandan, and S. Rajgopal (2021). Are carbon emissions associated with stock returns? *SSRN Working Paper 3800193*.

Attig, N., S. El Ghoul, O. Guedhami, and J. Suh (2013). Corporate social responsibility and credit ratings. *Journal of Business Ethics 117*(4), 679–694.

Atz, U., T. Van Holt, E. Douglas, and T. Whelan (2019). The return on sustainability investment (ROSI): Monetizing financial benefits of sustainability actions in companies. *Review of Business 39*(2), 1.

Auer, B. R. (2016). Do socially responsible investment policies add or destroy European stock portfolio value? *Journal of Business Ethics 135*(2), 381–397.

Auer, B. R. (2021). Implementation and profitability of sustainable investment strategies: An errors-in-variables perspective. *Business Ethics, the Environment & Responsibility 30*(4), 619–638.

Auer, B. R. and F. Schuhmacher (2016). Do socially (ir) responsible investments pay? New evidence from international ESG data. *Quarterly Review of Economics and Finance 59*, 51–62.

Austin, D. (2020). Milton friedman's hazardous feedback loop. *Responsible Investor*.

Avramov, D., S. Cheng, A. Lioui, and A. Tarelli (2021). Sustainable investing with esg rating uncertainty. *Journal of Financial Economics Forthcoming*.

Awaysheh, A., R. A. Heron, T. Perry, and J. I. Wilson (2020). On the relation between corporate social responsibility and financial performance. *Strategic Management Journal 41*(6), 965–987.

Azar, J., M. Duro, I. Kadach, and G. Ormazabal (2021). The big three and corporate carbon emissions around the world. *Journal of Financial Economics 142*(2), 674–696.

Azevedo, V., C. Kaserer, and L. MS Campos (2021). Investor sentiment and the time-varying sustainability premium. *Journal of Asset Management Forthcoming*.

Bae, J., Z. Sun, and L. Zheng (2019). Religious belief and socially responsible investing. *SSRN Working Paper 3414530*.

Bajic, S. and B. Yurtoglu (2018). CSR, market value and profitability: International evidence. In *Research handbook of finance and sustainability*. Edward Elgar Publishing.

Bams, D., B. van der Kroft, and K. Maas (2021). Heterogeneous CSR approaches, corporate social performance and corporate financial performance. *SSRN Working Paper 3906715*.

Bansal, R., D. A. Wu, and A. Yaron (2021). Is socially responsible investing a luxury good? *Review of Financial Studies Forthcoming*.

Barber, B. M., A. Morse, and A. Yasuda (2021). Impact investing. *Journal of Financial Economics 139*(1), 162–185.

Barko, T., M. Cremers, and L. Renneboog (2021). Shareholder engagement on environmental, social, and governance performance. *Journal of Business Ethics Forthcoming.*

Barnett, M. (2017). Climate change and uncertainty: An asset pricing perspective. *Working Paper.*

Barnett, M., W. Brock, and L. P. Hansen (2020). Pricing uncertainty induced by climate change. *Review of Financial Studies 33*(3), 1024–1066.

Barnett, M. L. and R. M. Salomon (2006). Beyond dichotomy: The curvilinear relationship between social responsibility and financial performance. *Strategic Management Journal 27*(11), 1101–1122.

Baron, D. P., M. A. Harjoto, and H. Jo (2011). The economics and politics of corporate social performance. *Business and Politics 13*(2), 1–46.

Barracchini, C. and M. E. Addessi (2012). Ethical portfolio theory: A new course. *Journal of Management & Sustainability 2*, 35–42.

Barreda-Tarrazona, I., J. C. Matallín-Sáez, and M. R. Balaguer-Franch (2011). Measuring investors' socially responsible preferences in mutual funds. *Journal of Business Ethics 103*(2), 305–330.

Barrios, J. M., M. Fasan, and D. Nanda (2014). Is corporate social responsibility an agency problem? evidence from ceo turnovers. *SSRN Working Paper 2540753.*

Barroso, J. S. S. and E. A. Araújo (2021). Socially responsible investments (sris)–mapping the research field. *Social Responsibility Journal 17*(4), 508–523.

Bartram, S. M., K. Hou, and S. Kim (2021). Real effects of climate policy: Financial constraints and spillovers. *Journal of Financial Economics Forthcoming.*

Basso, A. and S. Funari (2014). Constant and variable returns to scale DEA models for socially responsible investment funds. *European Journal of Operational Research 235*(3), 775–783.

Bătae, O. M., V. D. Dragomir, and L. Feleagă (2021). The relationship between Environmental, Social, and Financial performance in the banking sector: A European study. *Journal of Cleaner Production 290*, 125791.

Battiston, S., Y. Dafermos, and I. Monasterolo (2021). Climate risks and financial stability. *Journal of Financial Stability 54*, 100867.

Battiston, S., A. Mandel, I. Monasterolo, F. Schütze, and G. Visentin (2017). A climate stress-test of the financial system. *Nature Climate Change 7*(4), 283–288.

Bauer, R., K. Koedijk, and R. Otten (2005). International evidence on ethical mutual fund performance and investment style. *Journal of Banking & Finance 29*(7), 1751–1767.

Bauer, R., T. Ruof, and P. Smeets (2021). Get real! Individuals prefer more sustainable investments. *Review of Financial Studies 34*(8), 3976–4043.

Bax, K., Ö. Sahin, C. Czado, and S. Paterlini (2021). ESG, risk, and (tail) dependence. *SSRN Working Paper 3846739.*

Beal, D. J., M. Goyen, and P. Philips (2005). Why do we invest ethically? *Journal of Investing 14*(3), 66–78.

Bebchuk, L., A. Cohen, and A. Ferrell (2009). What matters in corporate governance? *Review of Financial Studies 22*(2), 783–827.

Bebchuk, L. and S. Hirst (2019). Index funds and the future of corporate governance. *Columbia Law Review 119*(8), 2029–2146.

Bebchuk, L. A. and A. Cohen (2005). The costs of entrenched boards. *Journal of Financial Economics 78*(2), 409–433.

Bebchuk, L. A., A. Cohen, and C. C. Wang (2013). Learning and the disappearing association between governance and returns. *Journal of Financial Economics 108*(2), 323–348.

Bebchuk, L. A. and R. Tallarita (2020). The illusory promise of stakeholder governance. *Cornell L. Rev. 106*, 91.

Becchetti, L., R. Ciciretti, and A. Dalò (2018). Fishing the corporate social responsibility risk factors. *Journal of Financial Stability 37*, 25–48.

Becchetti, L., R. Ciciretti, A. Dalò, and S. Herzel (2015). Socially responsible and conventional investment funds: Performance comparison and the global financial crisis. *Applied Economics 47*(25), 2541–2562.

Behnam, R., D. Gillers, B. Litterman, L. Martinez-Diaz, J. M. Keenan, and S. Moch (2020). Managing climate risk in the US financial system. Technical report, Report of the Climate-Related Market Risk Subcommittee, Market Risk Advisory Committee of the U.S. Commodity Futures Trading Commission.

Belghitar, Y., E. Clark, and N. Deshmukh (2017). Importance of the fund management company in the performance of socially responsible mutual funds. *Journal of Financial Research 40*(3), 349–367.

Bello, Z. Y. (2005). Socially responsible investing and portfolio diversification. *Journal of Financial Research 28*(1), 41–57.

Bender, J., T. A. Bridges, C. He, A. Lester, and X. Sun (2018). A blueprint for integrating ESG into equity portfolios. *Journal of Investment Management 16*(1).

Bender, J., C. He, C. Ooi, and X. Sun (2020). Reducing the carbon intensity of low volatility portfolios. *Journal of Portfolio Management 46*(3), 108–122.

Bender, J., X. Sun, and T. Wang (2017). Thematic indexing, meet smart beta! Merging ESG into factor portfolios. *Journal of Index Investing 8*(3), 89–101.

Benedetti, D., E. Biffis, F. Chatzimichalakis, L. L. Fedele, and I. Simm (2021). Climate change investment risk: Optimal portfolio construction ahead of the transition to a lower-carbon economy. *Annals of Operations Research 299*, 847–871.

Bengo, I., A. Borrello, and V. Chiodo (2021). Preserving the integrity of social impact investing: Towards a distinctive implementation strategy. *Sustainability 13*(5), 2852.

Benlemlih, M. (2019). Corporate social responsibility and dividend policy. *Research in International Business and Finance 47*, 114–138.

Benmir, G., I. Jaccard, and G. Vermandel (2020). Green asset pricing. *SSRN Working Paper 3706133*.

Bennani, L., T. Le Guenedal, F. Lepetit, L. Ly, V. Mortier, T. Roncalli, and T. Sekine (2018). How ESG investing has impacted the asset pricing in the equity market. *SSRN Working Paper 3316862*.

Bennedsen, M., E. Hillebrand, and S. J. Koopman (2021). Modeling, forecasting, and nowcasting US CO_2 emissions using many macroeconomic predictors. *Energy Economics*, 105118.

Benson, K. L. and J. E. Humphrey (2008). Socially responsible investment funds: Investor reaction to current and past returns. *Journal of Banking & Finance 32*(9), 1850–1859.

Bento, N. and G. Gianfrate (2020). Determinants of internal carbon pricing. *Energy Policy 143*, 111499.

Berg, F., K. Fabisik, and Z. Sautner (2021). Is history repeating itself? The (un)predictable past of ESG ratings. *SSRN Working Paper 3722087*.

Berg, F., J. F. Koelbel, and R. Rigobon (2020). Aggregate confusion: The divergence of ESG ratings. *SSRN Working Paper 3438533*.

Berk, J. and J. H. van Binsbergen (2021). The impact of impact investing. *SSRN Working Paper 3909166*.

Berkovitch, J., D. Israeli, A. Rakshit, and S. A. Sridharan (2021). Does CSR engender trust? Evidence from investor reactions to corporate disclosures. *SSRN Working Paper 3858135*.

Bernal, O., M. Hudon, and F.-X. Ledru (2021). Are impact and financial returns mutually exclusive? evidence from publicly-listed impact investments. *Quarterly Review of Economics and Finance 81*, 93–112.

Bernile, G., V. Bhagwat, and S. Yonker (2018). Board diversity, firm risk, and corporate policies. *Journal of Financial Economics 127*(3), 588–612.

Bernstein, A., M. T. Gustafson, and R. Lewis (2019). Disaster on the horizon: The price effect of sea level rise. *Journal of Financial Economics 134*(2), 253–272.

Berry, T. C. and J. C. Junkus (2013). Socially responsible investing: An investor perspective. *Journal of Business Ethics 112*(4), 707–720.

Bertolotti, A. and M. Kent (2019). Carbon beta-a framework for determining carbon price impacts on valuation. *SSRN Working Paper 3446914*.

Bessec, M. and J. Fouquau (2020). Green sentiment in financial markets: A global warning. *SSRN Working Paper 3710489*.

Bianchi, R. J., R. Copp, M. L. Kremmer, and E. Roca (2010). Should funds invest in socially responsible investments during downturns? *Accounting Research Journal 23*(3).

Bilbao-Terol, A., M. Arenas-Parra, and V. Cañal-Fernández (2012a). A fuzzy multi-objective approach for sustainable investments. *Expert Systems with Applications 39*(12), 10904–10915.

Bilbao-Terol, A., M. Arenas-Parra, and V. Cañal-Fernández (2012b). Selection of socially responsible portfolios using goal programming and fuzzy technology. *Information Sciences 189*, 110–125.

Bilbao-Terol, A., M. Arenas-Parra, V. Cañal-Fernández, and C. Bilbao-Terol (2016). Multi-criteria decision making for choosing socially responsible investment within a behavioral portfolio theory framework: A new way of investing into a crisis environment. *Annals of Operations Research 247*(2), 549–580.

Billio, M., M. Costola, I. Hristova, C. Latino, and L. Pelizzon (2021). Inside the ESG ratings:(Dis) agreement and performance. *Corporate Social Responsibility and Environmental Management 28*(5), 1426–1445.

Bilsback, K. R., D. Kerry, B. Croft, B. Ford, S. H. Jathar, E. Carter, R. V. Martin, and J. R. Pierce (2020). Beyond SOx reductions from shipping: assessing the impact of NOx and carbonaceous-particle controls on human health and climate. *Environmental Research Letters 15*(12), 124046.

Bingler, J. A., M. Kraus, and M. Leippold (2021). Cheap talk and cherry-picking: What climatebert has to say on corporate climate risk disclosures. *SSRN Working Paper*.

Blankenberg, A.-K. and J. F. Gottschalk (2018). Is socially responsible investing (SRI) in stocks a competitive capital investment? A comparative analysis based on the performance of sustainable stocks. *SSRN Working Paper 3186094*.

Blitz, D. and F. J. Fabozzi (2017). Sin stocks revisited: Resolving the sin stock anomaly. *Journal of Portfolio Management 44*(1), 105–111.

Blitz, D. and L. Swinkels (2020). Is exclusion effective? *Journal of Portfolio Management 46*(3), 42–48.

Blitz, D. and L. Swinkels (2021a). Does excluding sin stocks cost performance? *Journal of Sustainable Finance & Investment Forthcoming*.

Blitz, D. and L. Swinkels (2021b). Who owns tobacco stocks? *SSRN Working Paper 3841450*.

Blitz, D., L. Swinkels, and J. A. van Zanten (2021). Does sustainable investing deprive unsustainable firms from fresh capital? *Journal of Impact and ESG Investing Forthcoming*.

Block, J. H., M. Hirschmann, and C. Fisch (2020). Which criteria matter when impact investors screen social enterprises? *Journal of Corporate Finance 66*, 101813.

Board, F. S. (2021). Fsb roadmap for addressing climate-related financial risks. *FSB Working Paper*.

Bofinger, Y., K. J. Heyden, B. Rock, and C. Bannier (2021). The sustainability trap: Active fund managers between ESG investing and fund overpricing. *Finance Research Letters Forthcoming*, 102160.

Bøhren, Ø. and S. Staubo (2016). Mandatory gender balance and board independence. *European Financial Management 22*(1), 3–30.

Bollen, N. P. (2007). Mutual fund attributes and investor behavior. *Journal of Financial and Quantitative Analysis 42*(3), 683–708.

Bolliger, G. and D. Cornilly (2021). Sustainability attribution: The case of carbon intensity. *Journal of Impact & ESG Investing 2*(1), 93–99.

Bolton, P. and M. T. Kacperczyk (2020). Carbon premium around the world. *SSRN Working Paper 3550233.*

Bolton, P. and M. T. Kacperczyk (2021). Do investors care about carbon risk? *Journal of Financial Economics 142*, 517–549.

Bolton, P., T. Li, E. Ravina, and H. Rosenthal (2020). Investor ideology. *Journal of Financial Economics 137*(2), 320–352.

Borgers, A., J. Derwall, K. Koedijk, and J. Ter Horst (2013). Stakeholder relations and stock returns: On errors in investors' expectations and learning. *Journal of Empirical Finance 22*, 159–175.

Borghei, Z. (2021). Carbon disclosure: a systematic literature review. *Accounting & Finance Forthcoming.*

Borms, S., K. Boudt, F. Van Holle, and J. Willems (2021). Semi-supervised text mining for monitoring the news about the ESG performance of companies. In *Data Science for Economics and Finance*, pp. 217–239. Springer, Cham.

Bose, S. (2020). Evolution of ESG reporting frameworks. In *Values at Work*, pp. 13–33. Springer.

Bosquet, B. (2000). Environmental tax reform: does it work? A survey of the empirical evidence. *Ecological Economics 34*(1), 19–32.

Boubaker, S., L. Chourou, D. Himick, and S. Saadi (2017). It's about time! The influence of institutional investment horizon on corporate social responsibility. *Thunderbird International Business Review 59*(5), 571–594.

Boubaker, S., D. Cumming, and D. K. Nguyen (2018). *Research handbook of finance and sustainability*. Edward Elgar Publishing.

Bouye, E. and D. D. Menville (2021). The convergence of sovereign environmental, social and governance ratings. *SSRN Working Paper 3568547*.

Brakman Reiser, D. and A. M. Tucker (2020). Buyer beware: Variation and opacity in ESG and ESG index funds. *Cardozo Law Review 41*.

Brammer, S., C. Brooks, and S. Pavelin (2006). Corporate social performance and stock returns: UK evidence from disaggregate measures. *Financial Management 35*(3), 97–116.

Brammer, S. and A. Millington (2008). Does it pay to be different? An analysis of the relationship between corporate social and financial performance. *Strategic Management Journal 29*(12), 1325–1343.

Branch, B. and L. Cai (2012). Do socially responsible index investors incur an opportunity cost? *Financial Review 47*(3), 617–630.

Branch, M., L. R. Goldberg, and P. Hand (2019). A guide to ESG portfolio construction. *Journal of Portfolio Management 45*(4), 61–66.

Brandon, R. G., S. Glossner, P. Krueger, P. Matos, and T. Steffen (2021). Do responsible investors invest responsibly? *SSRN Working Paper 3525530*.

Brav, A. and J. Heaton (2021). Brown assets for the prudent investor. *SSRN Working Paper 3895887*.

Breedt, A., S. Ciliberti, S. Gualdi, and P. Seager (2019). Is ESG an equity factor or just an investment guide? *Journal of Investing 28*(2), 32–42.

Breitenstein, M., D. K. Nguyen, and T. Walther (2021). Environmental hazards and risk management in the financial sector: A systematic literature review. *Journal of Economic Surveys 35*(2), 512–538.

Briere, M. and S. Ramelli (2021a). Green sentiment, stock returns, and corporate behavior. *SSRN Working Paper 3850923*.

Briere, M. and S. Ramelli (2021b). Responsible investing and stock allocation. *SSRN Working Paper 3853256*.

Bril, H., G. Kell, and A. Rasche (2020). *Sustainable Investing: A Path to a New Horizon.* Routledge.

Brøgger, A. and A. Kronies (2020). Skills and sentiment in sustainable investing. *SSRN Working Paper 3531312*.

Brooks, C. and I. Oikonomou (2018). The effects of environmental, social and governance disclosures and performance on firm value: A review of the literature in accounting and finance. *British Accounting Review 50*(1), 1–15.

Brown, J. R., G. Martinsson, and C. J. Thomann (2021). Can environmental policy encourage technical change? emissions taxes and R&D investment in polluting firms. *SSRN Working Paper 3871447*.

Bruder, B., Y. Cheikh, F. Deixonne, and B. Zheng (2019). Integration of ESG in asset allocation. *SSRN Working Paper 3473874*.

Bruno, G., M. Esakia, and F. Goltz (2021). "Honey, i shrunk the ESG alpha": Risk-adjusting ESG portfolio returns. *Scientific Beta Working Paper*.

Bryant, P. (2020). *Sustainable Investing for Everyone: A Private Investor's Guide.* Canbry Consulting Limited.

Brzeszczynski, J., B. Ghimire, T. Jamasb, and G. McIntosh (2019). Socially responsible investment and market performance: the case of energy and resource companies. *Energy Journal 40*(5), 17–72.

Bu, L., K. C. Chan, A. Choi, and G. Zhou (2021). Talented inside directors and corporate social responsibility: A tale of two roles. *Journal of Corporate Finance 70*, 102044.

Burbano, V. C. (2021). Getting gig workers to do more by doing good: Field experimental evidence from online platform labor marketplaces. *Organization & Environment 34*(3), 387–412.

Burghof, H.-P. and M. Gehrung (2021). Investing ethical: Harder than you think. *SSRN Working Paper 3783746*.

Burke, J. J. (2021). Do boards take environmental, social, and governance issues seriously? Evidence from media coverage and CEO dismissals. *Journal of Business Ethics Forthcoming.*

Busch, T., A. Bassen, S. Lewandowski, and F. Sump (2020). Corporate carbon and financial performance revisited. *Organization & Environment Forthcoming.*

Cahan, S. F., C. Chen, L. Chen, and N. H. Nguyen (2015). Corporate social responsibility and media coverage. *Journal of Banking & Finance 59*, 409–422.

Cai, L. and C. He (2014). Corporate environmental responsibility and equity prices. *Journal of Business Ethics 125*(4), 617–635.

Cai, Y., H. Jo, and C. Pan (2012). Doing well while doing bad? CSR in controversial industry sectors. *Journal of Business Ethics 108*(4), 467–480.

Caiazza, S., G. Galloppo, and V. Paimanova (2021). The role of sustainability performance after merger and acquisition deals in short and long-term. *Journal of Cleaner Production 314*, 127982.

Cakici, N. and A. Zaremba (2021). Responsible investing: ESG ratings and the cross-section of international stock returns. *SSRN Working Paper 3922312*.

Calvo, C., C. Ivorra, and V. Liern (2015). Finding socially responsible portfolios close to conventional ones. *International Review of Financial Analysis 40*, 52–63.

Camejo, P. and G. Aiyer (2002). *The SRI advantage: Why socially responsible investing has outperformed financially.* New Society Pub.

Camilleri, M. A. (2020). The market for socially responsible investing: A review of the developments. *Social Responsibility Journal 17*(3), 412–428.

Cantino, V., A. Devalle, and S. Fiandrino (2017). ESG sustainability and financial capital structure: Where they stand nowadays. *International Journal of Business and Social Science 8*(5).

Cao, J., S. Titman, X. Zhan, and W. E. Zhang (2020). ESG preference, institutional trading, and stock return patterns. *SSRN Working Paper 3353623*.

Cao, S., W. Jiang, B. Yang, A. L. Zhang, et al. (2020). How to talk when a machine is listening: Corporate disclosure in the age of AI. *SSRN Working Paper 3683802*.

Capelle-Blancard, G., A. Desroziers, and O. D. Zerbib (2021). Socially responsible investing strategies under pressure: Evidence from COVID-19. *Journal of Portfolio Management 47*(9), 178–197.

Capelle-Blancard, G. and S. Monjon (2014). The performance of socially responsible funds: does the screening process matter? *European Financial Management 20*(3), 494–520.

Capelle-Blancard, G. and A. Petit (2019). Every little helps? ESG news and stock market reaction. *Journal of Business Ethics 157*(2), 543–565.

Cappucci, M. (2018). The ESG integration paradox. *Journal of Applied Corporate Finance 30*(2), 22–28.

Caragnano, A., M. Mariani, F. Pizzutilo, and M. Zito (2020). Is it worth reducing GHG emissions? Exploring the effect on the cost of debt financing. *Journal of Environmental Management 270*, 110860.

Carattini, S., S. Fankhauser, J. Gao, C. Gennaioli, and P. Panzarasa (2021). What does network analysis teach us about international environmental cooperation? *arXiv Preprint* (2106.08883).

Carattini, S. and S. Sen (2019). Carbon taxes and stranded assets: Evidence from Washington state. *SSRN Working Paper 3434841*.

Cash, D. (2018). Sustainable finance ratings as the latest symptom of 'rating addiction'. *Journal of Sustainable Finance & Investment 8*(3), 242–258.

Ceccarelli, M., S. Ramelli, and A. F. Wagner (2020). Low-carbon mutual funds. *SSRN Working Paper 3353239*.

Celiker, U., G. Sonaer, et al. (2021). Undervaluation of employee satisfaction. *Journal of Investing 30*(4), 51–64.

Chakrabarti, G. and C. Sen (2020). Time series momentum trading in green stocks. *Studies in Economics and Finance 37*(2), 361–389.

Chan, Y., K. Hogan, K. Schwaiger, and A. Ang (2020). ESG in factors. *Journal of Impact and ESG Investing 1*(1), 26–45.

Chang, C. E. and H. D. Witte (2010). Performance evaluation of US socially responsible mutual funds: Revisiting doing good and doing well. *American Journal of Business 25*(1), 9–21.

Chang, C.-L., J. Ilomäki, H. Laurila, and M. McAleer (2020). Causality between CO_2 emissions and stock markets. *Energies 13*(11), 2893.

Chatterji, A. K., R. Durand, D. I. Levine, and S. Touboul (2016). Do ratings of firms converge? Implications for managers, investors and strategy researchers. *Strategic Management Journal 37*(8), 1597–1614.

Chatterji, A. K., D. I. Levine, and M. W. Toffel (2009). How well do social ratings actually measure corporate social responsibility? *Journal of Economics & Management Strategy 18*(1), 125–169.

Chatzitheodorou, K., A. Skouloudis, K. Evangelinos, and I. Nikolaou (2019). Exploring socially responsible investment perspectives: A literature mapping and an investor classification. *Sustainable Production and Consumption 19*, 117–129.

Chaudhry, S. M., A. Saeed, and R. Ahmed (2021). Carbon neutrality: The role of banks in optimal environmental management strategies. *Journal of Environmental Management 299*, 113545.

Chava, S. (2014). Environmental externalities and cost of capital. *Management Science 60*(9), 2223–2247.

Chava, S., J. H. J. Kim, and J. Lee (2021). Doing well by doing good? risk, return, and environmental and social ratings. *SSRN Working Paper 3814444*.

Cheema-Fox, A., B. R. LaPerla, G. Serafeim, D. Turkington, and H. Wang (2021). Decarbonizing everything. *Financial Analysts Journal 77*(3), 93–108.

Cheema-Fox, A., B. R. LaPerla, G. Serafeim, D. Turkington, and H. S. Wang (2019). Decarbonization factors. *SSRN Working Paper 3448637*.

Cheema-Fox, A., G. Serafeim, and H. S. Wang (2021). Climate change vulnerability and currency returns. *SSRN Working Paper 3864393*.

Chegut, A., H. Schenk, and B. Scholtens (2011). Assessing sri fund performance research: Best practices in empirical analysis. *Sustainable Development 19*(2), 77–94.

Chen, H. A., K. Karim, and A. Tao (2021). The effect of suppliers' corporate social responsibility concerns on customers' stock price crash risk. *Advances in Accounting 52*, 100516.

Chen, J., G. Tang, J. Yao, and G. Zhou (2020). Employee sentiment and stock returns. *SSRN Working Paper 3612922*.

Chen, L., Y. Chen, A. Kumar, and W. S. Leung (2020). Firm-level ESG news and active fund management. *SSRN Working Paper 3747085*.

Chen, L., L. Zhang, J. Huang, H. Xiao, and Z. Zhou (2021). Social responsibility portfolio optimization incorporating ESG criteria. *Journal of Management Science and Engineering 6*(1), 75–85.

Chen, M. and G. Mussalli (2020). An integrated approach to quantitative ESG investing. *Journal of Portfolio Management 46*(3), 65–74.

Chen, M., R. von Behren, and G. Mussalli (2021). The unreasonable attractiveness of more ESG data. *Journal of Portfolio Management Forthcoming*.

Chen, X. and B. Scholtens (2018). The urge to act: A comparison of active and passive socially responsible investment funds in the United States. *Corporate Social Responsibility and Environmental Management 25*(6), 1154–1173.

Chen, Y., A. Kumar, and C. Zhang (2019). Social sentiment and asset prices. *SSRN Working Paper 3331866*.

Cheung, A. W. (2011). Do stock investors value corporate sustainability? Evidence from an event study. *Journal of Business Ethics 99*(2), 145–165.

Cheung, A. W., M. Hu, and J. Schwiebert (2018). Corporate social responsibility and dividend policy. *Accounting & Finance 58*(3), 787–816.

Cheung, A. W. and W. C. Pok (2019). Corporate social responsibility and provision of trade credit. *Journal of Contemporary Accounting & Economics 15*(3), 100159.

Chew, S. H. and K. K. Li (2021). The moral investor: Sin stock aversion and virtue stock affinity. *SSRN Working Paper 3773971*.

Chiappini, H., G. Vento, and L. De Palma (2021). The impact of COVID-19 lockdowns on sustainable indexes. *Sustainability 13*(4), 1846.

Chintrakarn, P., P. Jiraporn, S. Tong, N. Jiraporn, and R. Proctor (2020). How do independent directors view corporate social responsibility (CSR)? Evidence from a quasi-natural experiment. *Financial Review 55*(4), 697–716.

Choi, D., Z. Gao, and W. Jiang (2020). Attention to global warming. *Review of Financial Studies 33*(3), 1112–1145.

Chou, C. and S. O. Kimbrough (2019). Talking about climate change: What are enterprises saying in their SEC filings? *SSRN Working Paper 3509765*.

Chouaibi, S., Y. Chouaibi, and G. Zouari (2021). Board characteristics and integrated reporting quality: evidence from ESG European companies. *EuroMed Journal of Business Forthcoming*.

Christensen, D. M., G. Serafeim, and A. Sikochi (2021). Why is corporate virtue in the eye of the beholder? The case of ESG ratings. *Accounting Review Forthcoming*.

Christiansen, L., H. Lin, J. Pereira, P. B. Topalova, and R. Turk-Ariss (2016). Gender diversity in senior positions and firm performance: Evidence from Europe. *SSRN Working Paper 2759759*.

Chu, L., B. Zhang, R. Zhang, and G. Zhou (2021). ESG and the market return. *SSRN Working Paper 3869272*.

Ciciretti, R., A. Dalò, and L. Dam (2020). The contributions of betas versus characteristics to the ESG premium. *SSRN Working Paper 3010234*.

Clark, G. L. and M. Viehs (2014). The implications of corporate social responsibility for investors: An overview and evaluation of the existing CSR literature. *SSRN Working Paper 2481877*.

Clarkson, P. M., Y. Li, and G. D. Richardson (2004). The market valuation of environmental capital expenditures by pulp and paper companies. *Accounting Review 79*(2), 329–353.

Climent, F. and P. Soriano (2011). Green and good? The investment performance of US environmental mutual funds. *Journal of Business Ethics 103*(2), 275–287.

Cochrane, J. H. (2009). *Asset pricing: Revised edition*. Princeton University Press.

Coffey, B. S. and G. E. Fryxell (1991). Institutional ownership of stock and dimensions of corporate social performance: An empirical examination. *Journal of Business Ethics 10*(6), 437–444.

Cohen, L., U. G. Gurun, and Q. H. Nguyen (2020). The ESG-innovation disconnect: Evidence from green patenting. *SSRN Working Paper 3718682*.

Cojoianu, T., A. G. Hoepner, and Y. Lin (2021). Impact vs. ESG investing in private markets. *SSRN Working Paper 3615094*.

Colacito, R., B. Hoffmann, and T. Phan (2019). Temperature and growth: A panel analysis of the United States. *Journal of Money, Credit and Banking 51*(2-3), 313–368.

Colak, G., T. Korkeamaki, and N. Meyer (2020). ESG and CEO turnover. *SSRN Working Paper 3710538*.

Conen, R. and S. Hartmann (2019). The hidden risks of ESG conformity-benefiting from the ESG life cycle. *SSRN Working Paper 3426204*.

Coqueret, G., S. Stiernegrip, C. Morgenstern, J. Kelly, J. Frey-Skött, and B. Österberg (2021). Boosting ESG-based optimization with asset pricing characteristics. *SSRN Working Paper 3877242*.

Core, J. E., W. R. Guay, and T. O. Rusticus (2006). Does weak governance cause weak stock returns? An examination of firm operating performance and investors' expectations. *Journal of Finance 61*(2), 655–687.

Core, J. E., R. W. Holthausen, and D. F. Larcker (1999). Corporate governance, chief executive officer compensation, and firm performance. *Journal of Financial Economics 51*(3), 371–406.

Cormier, D. and M. Magnan (1997). Investors' assessment of implicit environmental liabilities: An empirical investigation. *Journal of Accounting and Public Policy 16*(2), 215–241.

Cornell, B. and A. Damodaran (2020). Valuing ESG: Doing good or sounding good? *Journal of Impact and ESG Investing 1*(1), 76–93.

Cortez, M. C., F. Silva, and N. Areal (2012). Socially responsible investing in the global market: The performance of

US and European funds. *International Journal of Finance & Economics 17*(3), 254–271.

Cosemans, M., X. Hut, and M. van Dijk (2021). The impact of climate change on optimal asset allocation for long-term investors. *NETSPAR Working Paper*.

Costantini, V. and M. Mazzanti (2012). On the green and innovative side of trade competitiveness? The impact of environmental policies and innovation on EU exports. *Research Policy 41*(1), 132–153.

Cox, P., S. Brammer, and A. Millington (2004). An empirical examination of institutional investor preferences for corporate social performance. *Journal of Business Ethics 52*(1), 27–43.

Cramton, P., D. J. MacKay, A. Ockenfels, and S. Stoft (2017). *Global carbon pricing: the path to climate cooperation*. MIT Press.

Cremers, K. M., L. P. Litov, and S. M. Sepe (2017). Staggered boards and long-term firm value, revisited. *Journal of Financial Economics 126*(2), 422–444.

Cremers, K. M. and V. B. Nair (2005). Governance mechanisms and equity prices. *Journal of Finance 60*(6), 2859–2894.

Cremers, M. and A. Ferrell (2014). Thirty years of shareholder rights and firm valuation. *Journal of Finance 69*(3), 1167–1196.

Crifo, P., R. Durand, and J.-P. Gond (2019). Encouraging investors to enable corporate sustainability transitions: The case of responsible investment in France. *Organization & Environment 32*(2), 125–144.

Cui, B. and P. Docherty (2020). Stock price overreaction to ESG controversies. *SSRN Working Paper 3559915*.

Curtis, Q., J. E. Fisch, and A. Robertson (2021). Do ESG mutual funds deliver on their promises? *Michigan Law Review Forthcoming.*

Cyert, R. M. and J. G. March (1963). *A behavioral theory of the firm*, Volume 2. Englewood Cliffs, NJ.

Dafermos, Y., M. Nikolaidi, and G. Galanis (2018). Climate change, financial stability and monetary policy. *Ecological Economics 152*, 219–234.

Dai, R., R. Duan, H. Liang, and L. Ng (2021). Outsourcing climate change. *SSRN Working Paper 3765485.*

Dai, R., H. Liang, and L. Ng (2021). Socially responsible corporate customers. *Journal of Financial Economics 142*(2), 598–626.

Dai, W. and P. Meyer-Brauns (2021). Greenhouse gas emissions and expected returns. *SSRN Working Paper 3714874.*

Daniel, K. D., B. Litterman, and G. Wagner (2018). Applying asset pricing theory to calibrate the price of climate risk. *SSRN Working Paper* (2865533).

Daniels, L., Y. Stevens, and D. Pratt. Environmentally friendly and socially responsible investment in and by occupational pension funds in the USA and in the EU. *European Journal of Social Security 23*(3), 247–263.

D'Apice, V., G. Ferri, and M. Intonti (2021). Sustainable disclosure versus ESG intensity: Is there a cross effect between holding and SRI funds? *Corporate Social Responsibility and Environmental Management 28*(5), 1496–1510.

Dasgupta, P. (2008). Discounting climate change. *Journal of Risk and Uncertainty 37*(2-3), 141–169.

Daugaard, D. (2020). Emerging new themes in environmental, social and governance investing: a systematic literature review. *Accounting & Finance 60*(2), 1501–1530.

David, P., M. Bloom, and A. J. Hillman (2007). Investor activism, managerial responsiveness, and corporate social performance. *Strategic Management Journal 28*(1), 91–100.

Davis, T. and B. Lescott (2019). ESG as a value-creation tool for active investors: A profile of inherent group. *Journal of Applied Corporate Finance 31*(2), 42–49.

De, I. and M. R. Clayman (2015). The benefits of socially responsible investing: An active manager's perspective. *Journal of Investing 24*(4), 49–72.

de Franco, C. (2020). ESG controversies and their impact on performance. *Journal of Investing 29*(2), 33–45.

De Franco, C., J. Nicolle, and L.-A. Tran (2021). Sustainable investing: ESG versus SDG. *Ossiam Working Paper*.

de Groot, W., J. de Koning, and S. van Winkel (2021). Sustainable voting behavior of asset managers: Do they walk the walk? *Journal of Impact and ESG Investing 1*(4), 7–29.

De Haan, J. and R. Vlahu (2016). Corporate governance of banks: A survey. *Journal of Economic Surveys 30*(2), 228–277.

de la Fuente, G., M. Ortiz, and P. Velasco (2021). The value of a firm's engagement in ESG practices: Are we looking at the right side? *Long Range Planning Forthcoming*, 102143.

De Villiers, C., D. Ma, and A. Marques (2020). CSR disclosure, dividend pay-outs and firm value: Relations and mediating effects. *SSRN Working Paper 3700880*.

Delmas, M. A. and V. C. Burbano (2011). The drivers of greenwashing. *California Management Review 54*(1), 64–87.

Delmas, M. A., N. Nairn-Birch, and J. Lim (2015). Dynamics of environmental and financial performance: The case of greenhouse gas emissions. *Organization & Environment 28*(4), 374–393.

Demers, E., J. Hendrikse, P. Joos, and B. Lev (2021). ESG didn't immunize stocks against the Covid-19 market crash, but investments in intangible assets did. *Journal of Business Finance & Accounting 48*(3-4), 433–462.

Deng, X., J.-k. Kang, and B. S. Low (2013). Corporate social responsibility and stakeholder value maximization: Evidence from mergers. *Journal of Financial Economics 110*(1), 87–109.

Derchi, G.-B., L. Zoni, and A. Dossi (2021). Corporate social responsibility performance, incentives, and learning effects. *Journal of Business Ethics 173*, 617–641.

Derrien, F., P. Krueger, A. Landier, and T. Yao (2021). How do ESG incidents affect firm value? *SSRN Working Paper 3903274*.

Derwall, J., N. Guenster, R. Bauer, and K. Koedijk (2005). The eco-efficiency premium puzzle. *Financial Analysts Journal 61*(2), 51–63.

DesJardine, M. R., E. Marti, and R. Durand (2021). Why activist hedge funds target socially responsible firms: The reaction costs of signaling corporate social responsibility. *Academy of Management Journal 64*(3).

Dhaliwal, D. S., O. Z. Li, A. Tsang, and Y. G. Yang (2011). Voluntary nonfinancial disclosure and the cost of equity capital: The initiation of corporate social responsibility reporting. *Accounting Review 86*(1), 59–100.

Diaye, M.-A., S.-H. Ho, and R. Oueghlissi (2021). ESG performance and economic growth: a panel co-integration analysis. *Empirica Forthcoming*, 1–24.

Diaz, V., D. Ibrushi, and J. Zhao (2021). Reconsidering systematic factors during the Covid-19 pandemic–the rising importance of ESG. *Finance Research Letters 38*, 101870.

Dietrich, A., G. Müller, and R. Schoenle (2021). The expectations channel of climate change: Implications for monetary policy. *CEPR Working Paper 15866*.

Diez-Busto, E., L. Sanchez-Ruiz, and A. Fernandez-Laviada (2021). The B Corp movement: A systematic literature review. *Sustainability 13*(5), 2508.

Dikolli, S. S., M. M. Frank, M. Z. Guo, and L. J. Lynch (2021). Walk the talk: ESG mutual fund voting on shareholder proposals. *SSRN Working Paper 3849762*.

Dimson, E., O. Karakaş, and X. Li (2015). Active ownership. *Review of Financial Studies 28*(12), 3225–3268.

Dimson, E., P. Marsh, and M. Staunton (2020a). Divergent ESG ratings. *Journal of Portfolio Management 47*(1), 75–87.

Dimson, E., P. Marsh, and M. Staunton (2020b). Exclusionary screening. *Journal of Impact and ESG Investing 1*(1), 66–75.

Dinda, S. (2004). Environmental kuznets curve hypothesis: a survey. *Ecological Economics 49*(4), 431–455.

Dolvin, S., J. Fulkerson, and A. Krukover (2019). Do "good guys" finish last? The relationship between Morningstar sustainability ratings and mutual fund performance. *Journal of Investing 28*(2), 77–91.

Domini, A. L. (2001). *Socially responsible investing: Making a difference and making money.* Dearborn Trade Publishing.

Dorfleitner, G., C. Kreuzer, and R. Laschinger (2021). How socially irresponsible are socially responsible mutual funds?

A persistence analysis. *Finance Research Letters Forthcoming*, 101990.

Dorfleitner, G., C. Kreuzer, and C. Sparrer (2020). ESG controversies and controversial ESG: about silent saints and small sinners. *Journal of Asset Management 21*(5), 393–412.

Dorfleitner, G. and S. Utz (2012). Safety first portfolio choice based on financial and sustainability returns. *European Journal of Operational Research 221*(1), 155–164.

Döring, S., W. Drobetz, S. El Ghoul, O. Guedhami, and H. Schröder (2021). Institutional investor horizon and firm valuation around the world. *Journal of International Business Studies 52*, 212–244.

Dou, X., Y. Wang, P. Ciais, F. Chevallier, S. J. Davis, M. Crippa, G. Janssens-Maenhout, D. Guizzardi, E. Solazzo, F. Yan, et al. (2021). Global gridded daily CO_2 emissions. *arXiv Preprint* (2107.08586).

Drempetic, S., C. Klein, and B. Zwergel (2020). The influence of firm size on the ESG score: Corporate sustainability ratings under review. *Journal of Business Ethics 167*, 333–360.

Drupp, M. A., M. C. Freeman, B. Groom, and F. Nesje (2018). Discounting disentangled. *American Economic Journal: Economic Policy 10*(4), 109–34.

Drut, B. (2010). Social responsibility and mean-variance portfolio selection. *Working papers CEB 10*.

Du, X. (2015). How the market values greenwashing? Evidence from China. *Journal of Business Ethics 128*(3), 547–574.

Duanmu, J., Q. Huang, Y. Li, and G. McBrayer (2021). Can hedge funds benefit from CSR investment? *Financial Review 56*(2), 251–278.

Ducoulombier, F. (2021). Understanding the importance of scope 3 emissions and the implications of data limitations. *Journal of Impact and ESG Investing Forthcoming*.

Dunbar, C., Z. F. Li, and Y. Shi (2020). CEO risk-taking incentives and corporate social responsibility. *Journal of Corporate Finance 64*, 101714.

Dunbar, C. G., Z. F. Li, and Y. Shi (2021). Corporate social (ir) responsibility and firm risk: The role of corporate governance. *SSRN Working Paper 3791594*.

Dunz, N., A. Naqvi, and I. Monasterolo (2021). Climate sentiments, transition risk, and financial stability in a stock-flow consistent model. *Journal of Financial Stability 54*, 100872.

Dutordoir, M., N. C. Strong, and P. Sun (2018). Corporate social responsibility and seasoned equity offerings. *Journal of Corporate Finance 50*, 158–179.

Dyck, A., K. V. Lins, L. Roth, and H. F. Wagner (2019). Do institutional investors drive corporate social responsibility? International evidence. *Journal of Financial Economics 131*(3), 693–714.

Eccles, R. G., I. Ioannou, and G. Serafeim (2014). The impact of corporate sustainability on organizational processes and performance. *Management Science 60*(11), 2835–2857.

Eccles, R. G., M. D. Kastrapeli, and S. J. Potter (2017). How to integrate ESG into investment decision-making: Results of a global survey of institutional investors. *Journal of Applied Corporate Finance 29*(4), 125–133.

Edmans, A. (2011). Does the stock market fully value intangibles? employee satisfaction and equity prices. *Journal of Financial Economics 101*(3), 621–640.

Edmans, A. (2012). The link between job satisfaction and firm value, with implications for corporate social responsibility. *Academy of Management Perspectives 26*(4), 1–19.

Edmans, A. (2020). *Grow the Pie: How Great Companies Deliver Both Purpose and Profit.* Cambridge University Press.

Edmans, A., L. Li, and C. Zhang (2020). Employee satisfaction, labor market flexibility, and stock returns around the world. *SSRN Working Paper 2461003.*

El Ghoul, S., O. Guedhami, C. C. Kwok, and D. R. Mishra (2011). Does corporate social responsibility affect the cost of capital? *Journal of Banking & Finance 35*(9), 2388–2406.

El Ghoul, S., A. Karoui, S. Patel, and S. Ramani (2021). The green and brown performances of mutual fund portfolios. *SSRN Working Paper 3766404.*

Elmalt, D., D. Igan, and D. Kirti (2021). Limits to private climate change mitigation. *SSRN Working Paper 3846150.*

Endrikat, J., E. Guenther, and H. Hoppe (2014). Making sense of conflicting empirical findings: A meta-analytic review of the relationship between corporate environmental and financial performance. *European Management Journal 32*(5), 735–751.

Engle, R. F., S. Giglio, B. Kelly, H. Lee, and J. Stroebel (2020). Hedging climate change news. *Review of Financial Studies 33*(3), 1184–1216.

Erlandsson, U. (2021). Carbon negative leveraged investment strategies. *SSRN Working Paper 3906531.*

Escrig-Olmedo, E., M. Á. Fernández-Izquierdo, I. Ferrero-Ferrero, J. M. Rivera-Lirio, and M. J. Muñoz-Torres (2019). Rating the raters: Evaluating how ESG rating agencies integrate sustainability principles. *Sustainability 11*(3), 915.

Esty, D. C. and T. Cort (2020). *Values at Work: Sustainable Investing and ESG Reporting.* Springer Nature.

Evans, J. R. and D. Peiris (2010). The relationship between environmental social governance factors and stock returns. *SSRN Working Paper 1725077.*

Fabozzi, F. J., K. Ma, and B. J. Oliphant (2008). Sin stock returns. *Journal of Portfolio Management 35*(1), 82–94.

Faccini, R., R. Matin, and G. Skiadopoulos (2021). Dissecting climate risks: Are they reflected in stock prices? *SSRN Working Paper 3795964.*

Fairhurst, D. J. and D. Greene (2020). Too much of a good thing? Corporate social responsibility and the takeover market. *SSRN Working Paper 3374895.*

Falcão, S. M. F., R. A. R. Bezerra, S. G. R. da Luz, et al. (2020). Concepts and forms of greenwashing: A systematic review. *Environmental Sciences Europe 32*(1).

Fama, E. F. (2021). Contract costs, stakeholder capitalism, and ESG. *27*(2), 189–195.

Fama, E. F. and K. R. French (2015). A five-factor asset pricing model. *Journal of Financial Economics 116*(1), 1–22.

Fama, E. F. and J. D. MacBeth (1973). Risk, return, and equilibrium: Empirical tests. *Journal of Political Economy 81*(3), 607–636.

Fan, J. H. and L. Michalski (2020). Sustainable factor investing: Where doing well meets doing good. *International Review of Economics & Finance 70*, 230–256.

Fang, M., K. S. Tan, and T. S. Wirjanto (2019). Sustainable portfolio management under climate change. *Journal of Sustainable Finance & Investment 9*(1), 45–67.

Feng, M., X. Wang, and J. G. Kreuze (2017). Corporate social responsibility and firm financial performance. *American Journal of Business 32*(3/4).

Fernando, C., M. Sharfman, and V. Uysal (2010). Does greenness matter? The effect of corporate environmental performance on ownership structure, analyst coverage and firm value. *FMA European Conference.*

Fernando, C. S., M. P. Sharfman, and V. B. Uysal (2017). Corporate environmental policy and shareholder value: Following the smart money. *Journal of Financial and Quantitative Analysis 52*(5), 2023–2051.

Ferrell, A., H. Liang, and L. Renneboog (2016). Socially responsible firms. *Journal of Financial Economics 122*(3), 585–606.

Ferrés, D. and F. Marcet (2021). Corporate social responsibility and corporate misconduct. *Journal of Banking & Finance 127*, 106079.

Feuer, A. (2020). Ethics, earnings, and ERISA: Ethical-factor investing of savings and retirement benefits. *SSRN Working Paper 3740364.*

Fiaschi, D., E. Giuliani, F. Nieri, and N. Salvati (2020). How bad is your company? Measuring corporate wrongdoing beyond the magic of ESG metrics. *Business Horizons 63*, 287–299.

Filbeck, A., G. Filbeck, and X. Zhao (2019). Performance assessment of firms following Sustainalytics ESG principles. *Journal of Investing 28*(2), 7–20.

Filbeck, G., T. A. Krause, and L. Reis (2016). Socially responsible investing in hedge funds. *Journal of Asset Management 17*(6), 408–421.

Finkelstein Shapiro, A. and G. E. Metcalf (2021). The macroeconomic effects of a carbon tax to meet the US Paris agreement target: The role of firm creation and technology adoption. *SSRN Working Paper 3856914*.

Fiordelisi, F., G. Galloppo, G. Lattanzio, and V. Paimanova (2021). An ESG ratings free assessment of socially responsible investment strategies. *SSRN Working Paper 3743802*.

Fish, A., D. Kim, and S. Venkatraman (2019). The ESG sacrifice. *SSRN Working Paper 3488475*.

Fisher-Vanden, K. and K. S. Thorburn (2011). Voluntary corporate environmental initiatives and shareholder wealth. *Journal of Environmental Economics and Management 62*(3), 430–445.

Fleurbaey, M. and S. Zuber (2015). Discounting, risk and inequality: A general approach. *Journal of Public Economics 128*, 34–49.

Folger-Laronde, Z., S. Pashang, L. Feor, and A. ElAlfy (2020). ESG ratings and financial performance of exchange-traded funds during the covid-19 pandemic. *Journal of Sustainable Finance & Investment Forthcoming*, 1–7.

Francoeur, C., F. Lakhal, S. Gaaya, and I. B. Saad (2021). How do powerful CEOs influence corporate environmental performance? *Economic Modelling 94*, 121–129.

Freiberg, D., J. Grewal, and G. Serafeim (2021). Science-based carbon emissions targets. *SSRN Working Paper 3804530*.

Friede, G., T. Busch, and A. Bassen (2015). ESG and financial performance: aggregated evidence from more than 2000 empirical studies. *Journal of Sustainable Finance & Investment 5*(4), 210–233.

Friedman, M. (1970). A Friedman doctrine: The social responsibility of business is to increase its profits. *New York Times Magazine 13*(1970), 32–33.

Fu, X., Y. Lin, and Y. Zhang (2020). Responsible investing in the gaming industry. *Journal of Corporate Finance 64*, 101657.

Gaganis, C., F. Pasiouras, M. Tasiou, and C. Zopounidis (2021). CISEF: a Composite Index of Social, Environmental and Financial Performance. *European Journal of Operational Research 291*(1), 394–409.

Galema, R., A. Plantinga, and B. Scholtens (2008). The stocks at stake: Return and risk in socially responsible investment. *Journal of Banking & Finance 32*(12), 2646–2654.

Gantchev, N. and M. Giannetti (2020). The costs and benefits of shareholder democracy: Gadflies and low-cost activism. *Review of Financial Studies Forthcoming.*

Gantchev, N., M. Giannetti, and R. Li (2020). Sustainability or performance? ratings and fund managers' incentives. *SSRN Working Paper 3731006.*

Gao, L., J. He, and J. Wu (2021). Standing out from the crowd via corporate social responsibility: Evidence from non-fundamental-driven price pressure. *SSRN Working Paper 2830742.*

Gasser, S. M., M. Rammerstorfer, and K. Weinmayer (2017). Markowitz revisited: Social portfolio engineering. *European Journal of Operational Research 258*(3), 1181–1190.

Gaussel, N. and L. Le Saint (2020). ESG risk rating of alternative portfolios. *SSRN Working Paper 3721898.*

Geczy, C. and J. Guerard (2021). ESG and expected returns on equities: The case of environmental ratings. *SSRN Working Paper 3903480.*

Geczy, C., J. Guerard, and M. Samonov (2019). Efficient SRI / ESG portfolios. *SSRN Working Paper 3011644*.

Geczy, C., R. F. Stambaugh, and D. Levin (2021). Investing in socially responsible mutual funds. *Review of Asset Pricing Studies 11*(2), 309–351.

Gelrud, M. (2021). Discounting climate change mitigating projects: A production-based model with disasters. *SSRN Working Paper 3778926*.

Gerard, B. (2018). ESG and socially responsible investment: A critical review. *SSRN Working Paper 3309650*.

Gerged, A. M., E. Beddewela, and C. J. Cowton (2021). Is corporate environmental disclosure associated with firm value? A multicountry study of gulf cooperation council firms. *Business Strategy and the Environment 30*(1), 185–203.

Gerged, A. M., L. Matthews, and M. Elheddad (2021). Mandatory disclosure, greenhouse gas emissions and the cost of equity capital: UK evidence of a U-shaped relationship. *Business Strategy and the Environment 30*(2), 908–930.

Ghosh, C., M. T. Petrova, J. Sun, and Y. Xiao (2016). Increasing gender diversity in corporate boards: Are firms catering to investor preferences? *SSRN Working Paper 2870077*.

Gianfrate, G., T. Kievid, and M. van Dijk (2021). On the resilience of ESG stocks during COVID-19: Global evidence. *CEPR Press 83*, 25–53.

Gianfrate, G., D. Schoenmaker, and S. Wasama (2015). Cost of capital and sustainability: a literature review. *Rotterdam School of Management Working Paper*.

Gibson, R., S. Glossner, P. Krueger, P. Matos, and T. Steffen (2020). Responsible institutional investing around the world. *SSRN Working Paper 3525530*.

Gibson, R. and P. Krueger (2018). The sustainability footprint of institutional investors. *SSRN Working Paper 2918926*.

Gibson, R., P. Krueger, N. Riand, and P. S. Schmidt (2021). ESG rating disagreement and stock returns. *Financial Analysts Journal 77*(4).

Gidwani, B. (2020). Some issues with using ESG ratings in an investment process. *Journal of Investing 29*(6), 76–84.

Giese, G., L.-E. Lee, D. Melas, Z. Nagy, and L. Nishikawa (2019). Foundations of ESG investing: how ESG affects equity valuation, risk, and performance. *Journal of Portfolio Management 45*(5), 69–83.

Giese, G., Z. Nagy, and L.-E. Lee (2021). Deconstructing ESG ratings performance: Risk and return for E, S, and G by time horizon, sector, and weighting. *Journal of Portfolio Management Forthcoming*.

Giese, G., Z. Nagy, and B. Rauis (2021). Foundations of climate investing: How equity markets have priced climate-transition risks. *Journal of Portfolio Management Forthcoming*.

Giglio, S., B. Kelly, and J. Stroebel (2021). Climate finance. *Annual Review of Financial Economics 13*.

Gil-Bazo, J., P. Ruiz-Verdú, and A. A. Santos (2010). The performance of socially responsible mutual funds: The role of fees and management companies. *Journal of Business Ethics 94*(2), 243–263.

Gillan, S. L., A. Koch, and L. T. Starks (2021). Firms and social responsibility: A review of ESG and CSR research in corporate finance. *Journal of Corporate Finance 66*, 101889.

Gillan, S. L. and L. T. Starks (2000). Corporate governance proposals and shareholder activism: The role of institutional investors. *Journal of Financial Economics 57*(2), 275–305.

Gillan, S. L. and L. T. Starks (2007). The evolution of shareholder activism in the united states. *Journal of Applied Corporate Finance 19*(1), 55–73.

Giroud, X. and H. M. Mueller (2011). Corporate governance, product market competition, and equity prices. *Journal of Finance 66*(2), 563–600.

Glac, K. (2012). The impact and source of mental frames in socially responsible investing. *Journal of Behavioral Finance 13*(3), 184–198.

Gloßner, S. (2019). Investor horizons, long-term blockholders, and corporate social responsibility. *Journal of Banking & Finance 103*, 78–97.

Gloßner, S. (2021). ESG incidents and shareholder value. *SSRN Working Paper 3004689.*

Glossner, S., P. Matos, S. Ramelli, and A. F. Wagner (2021). Do institutional investors stabilize equity markets in crisis periods? evidence from covid-19. *SSRN Working Paper 3655271.*

Glück, M., B. Hübel, and H. Scholz (2021). ESG rating events and stock market reactions. *SSRN Working Paper 3803254.*

Goktan, M. S., R. Kieschnick, and R. Moussawi (2018). Corporate governance and firm survival. *Financial Review 53*(2), 209–253.

Goldberg, L. R. and S. Mouti (2019). Sustainable investing and the cross-section of maximum drawdown. *SSRN Working Paper 3387615.*

Gollier, C. (2002). Discounting an uncertain future. *Journal of Public Economics 85*(2), 149–166.

Gollier, C. (2008). Discounting with fat-tailed economic growth. *Journal of Risk and Uncertainty 37*(2-3), 171–186.

Gollier, C. (2013). *Pricing the planet's future: the economics of discounting in an uncertain world.* Princeton University Press.

Gollier, C. (2017). *Ethical Asset Valuation and the Good Society.* Columbia University Press.

Gollier, C. (2021). The cost-efficiency carbon pricing puzzle. *SSRN Working Paper 3805342.*

Gomes, M. and S. Marsat (2018). Does CSR impact premiums in M&A transactions? *Finance Research Letters 26*, 71–80.

Gompers, P., J. Ishii, and A. Metrick (2003). Corporate governance and equity prices. *Quarterly Journal of Economics 118*(1), 107–156.

Gomtsian, S. (2020). Shareholder engagement by large institutional investors. *Journal of Corporation Law Forthcoming.*

Gonçalves, T., D. Pimentel, and C. Gaio (2021). Risk and performance of European green and conventional funds. *Sustainability 13*(8), 4226.

Goodwin, S. (2016). The long-term efficacy of activist directors. *SSRN Working Paper 2731369.*

Görgen, M., A. Jacob, and M. Nerlinger (2021). Get green or die trying? Carbon risk integration into portfolio management. *Journal of Portfolio Management 47*(3), 77–93.

Görgen, M., A. Jacob, M. Nerlinger, R. Riordan, M. Rohleder, and M. Wilkens (2020). Carbon risk. *SSRN Working Paper 2930897.*

Gostlow, G. (2019). Pricing climate risk. *SSRN Working Paper 3501013.*

Gratcheva, E., T. Emery, and D. Wang (2021). Demystifying sovereign ESG. *SSRN Working Paper 3854177*.

Green, D. and B. N. Roth (2020). The allocation of socially responsible capital. *SSRN Working Paper 3737772*.

Green, J. F. (2021). Does carbon pricing reduce emissions? A review of ex-post analyses. *Environmental Research Letters 16*.

Gregory, A., R. Tharyan, and J. Whittaker (2014). Corporate social responsibility and firm value: Disaggregating the effects on cash flow, risk and growth. *Journal of Business Ethics 124*(4), 633–657.

Gregory, A. and J. Whittaker (2013). Exploring the valuation of corporate social responsibility—a comparison of research methods. *Journal of Business Ethics 116*(1), 1–20.

Grewal, J., C. Hauptmann, and G. Serafeim (2021). Material sustainability information and stock price informativeness. *Journal of Business Ethics 171*(3), 513–544.

Grewal, J. and G. Serafeim (2020). Research on corporate sustainability: Review and directions for future research. *Foundations and Trends® in Accounting 14*, 73–127.

Griffin, D. W., O. Guedhami, K. Li, and G. Lu (2021). National culture and the value implications of corporate social responsibility. *SSRN Working Paper 3250222*.

Griffin, J. J. and J. F. Mahon (1997). The corporate social performance and corporate financial performance debate: Twenty-five years of incomparable research. *Business & Society 36*(1), 5–31.

Grim, D. M. and D. B. Berkowitz (2020). ESG, SRI, and impact investing: A primer for decision-making. *Journal of Impact and ESG Investing 1*(1), 47–65.

Guenster, N., R. Bauer, J. Derwall, and K. Koedijk (2011). The economic value of corporate eco-efficiency. *European Financial Management 17*(4), 679–704.

Guerard Jr, J. B. (1997). Is there a cost to being socially responsible in investing? *Journal of Forecasting 16*(7), 475–490.

Guest, P. M. and M. Nerino (2020). Do corporate governance ratings change investor expectations? Evidence from announcements by Institutional Shareholder Services. *Review of Finance 24*(4), 891–928.

Gurvich, A. and G. G. Creamer (2021). Carbon risk factor framework. *SSRN Working Paper 3901956*.

Gyönyörová, L., M. Stachoň, and D. Stašek (2021). ESG ratings: relevant information or misleading clue? evidence from the S&P global 1200. *Journal of Sustainable Finance & Investment Forthcoming*, 1–35.

Hahn, R., D. Reimsbach, and F. Schiemann (2015). Organizations, climate change, and transparency: Reviewing the literature on carbon disclosure. *Organization & Environment 28*(1), 80–102.

Hain, L. I., J. F. Kölbel, and M. Leippold (2021). Let's get physical: Comparing metrics of physical climate risk. *Finance Research Letters Forthcoming*.

Halbritter, G. and G. Dorfleitner (2015). The wages of social responsibility—where are they? a critical review of ESG investing. *Review of Financial Economics 26*, 25–35.

Hallerbach, W., H. Ning, A. Soppe, and J. Spronk (2004). A framework for managing a portfolio of socially responsible investments. *European Journal of Operational Research 153*(2), 517–529.

Hamilton, S., H. Jo, and M. Statman (1993). Doing well while doing good? The investment performance of socially responsible mutual funds. *Financial Analysts Journal 49*(6), 62–66.

Hang, M., J. Geyer-Klingeberg, A. Rathgeber, and S. Stöckl (2018). Economic development matters: A meta-regression analysis on the relation between environmental management and financial performance. *Journal of Industrial Ecology 22*(4), 720–744.

Hansen, J., R. Ruedy, M. Sato, and K. Lo (2010). Global surface temperature change. *Reviews of Geophysics 48*(4).

Haque, F. (2017). The effects of board characteristics and sustainable compensation policy on carbon performance of UK firms. *British Accounting Review 49*(3), 347–364.

Haque, F. and C. G. Ntim (2018). Environmental policy, sustainable development, governance mechanisms and environmental performance. *Business Strategy and the Environment 27*(3), 415–435.

Haque, F. and C. G. Ntim (2020). Executive compensation, sustainable compensation policy, carbon performance and market value. *British Journal of Management 31*(3), 525–546.

Harjoto, M., H. Jo, and Y. Kim (2017). Is institutional ownership related to corporate social responsibility? The nonlinear relation and its implication for stock return volatility. *Journal of Business Ethics 146*(1), 77–109.

Harper, H. (2020). One institutional investor's approach to integrating ESG in the investment process. *Journal of Portfolio Management 46*(4), 110–123.

Harper Ho, V. E. (2021). Modernizing ESG disclosure. *University of Illinois Law Review Forthcoming*.

Hart, O. and L. Zingales (2017). Companies should maximize shareholder welfare not market value. *Journal of Law, Finance, and Accounting 2*, 247–274.

Hart, S. L. and G. Ahuja (1996). Does it pay to be green? An empirical examination of the relationship between emission reduction and firm performance. *Business Strategy and the Environment 5*(1), 30–37.

Hartzmark, S. M. and A. B. Sussman (2019). Do investors value sustainability? A natural experiment examining ranking and fund flows. *Journal of Finance 74*(6), 2789–2837.

Hasan, I., N. Kobeissi, L. Liu, and H. Wang (2018). Corporate social responsibility and firm financial performance: The mediating role of productivity. *Journal of Business Ethics 149*(3), 671–688.

He, J. and P. Richard (2010). Environmental Kuznets curve for CO_2 in canada. *Ecological Economics 69*(5), 1083–1093.

He, Y., B. Kahraman, and M. Lowry (2020). ES risks and shareholder voice. *SSRN Working Paper 3284683*.

He, Y., Q. Tang, and K. Wang (2013). Carbon disclosure, carbon performance, and cost of capital. *China Journal of Accounting Studies 1*(3-4), 190–220.

Heal, G. (2009). Climate economics: a meta-review and some suggestions for future research. *Review of Environmental Economics and Policy 3*(1), 4–21.

Heath, D., D. Macciocchi, R. Michaely, and M. C. Ringgenberg (2021). Does socially responsible investing change firm behavior? *SSRN Working Paper 3837706*.

Heaton, J. (2021). Expected returns to brown and green assets with transition uncertainty. *SSRN Working Paper 3910042*.

Hedblom, D., B. R. Hickman, and J. A. List (2021). Toward an understanding of corporate social responsibility: Theory and field experimental evidence. *SSRN Working Paper 3450248*.

Heeb, F., J. F. Kölbel, F. Paetzold, and S. Zeisberger (2021). Do investors care about impact? *SSRN Working Paper 3765659*.

Hellström, J., N. Lapanan, and R. Olsson (2020). Socially responsible investments among parents and adult children. *European Economic Review 121*, 103328.

Henriksson, R., J. Livnat, P. Pfeifer, and M. Stumpp (2019). Integrating ESG in portfolio construction. *Journal of Portfolio Management 45*(4), 67–81.

Heo, Y. (2021). Climate change exposure and firm cash holdings. *SSRN Working Paper 3795298*.

Herremans, I. M., P. Akathaporn, and M. McInnes (1993). An investigation of corporate social responsibility reputation and economic performance. *Accounting, Organizations and Society 18*(7-8), 587–604.

Hilario-Caballero, A., A. Garcia-Bernabeu, J. V. Salcedo, and M. Vercher (2020). Tri-criterion model for constructing low-carbon mutual fund portfolios: a preference-based multi-objective genetic algorithm approach. *International Journal of Environmental Research and Public Health 17*(17), 6324.

Hill, J. (2020). *Environmental, Social, and Governance (ESG) investing: A balanced analysis of the theory and practice of a sustainable portfolio.* Academic Press.

Hill, R. P., T. Ainscough, T. Shank, and D. Manullang (2007). Corporate social responsibility and socially responsible investing: A global perspective. *Journal of Business Ethics 70*(2), 165–174.

Hillman, A. J. and G. D. Keim (2001). Shareholder value, stakeholder management, and social issues: What's the bottom line? *Strategic Management Journal 22*(2), 125–139.

Hjort, I. (2016). Potential climate risks in financial markets: A literature overview. *Working Paper from the Economics Department of the University of Oslo.*

Hoang, T. H. V., W. Przychodzen, J. Przychodzen, and E. A. Segbotangni (2021). Environmental transparency and performance: Does the corporate governance matter? *Environmental and Sustainability Indicators 10*, 100123.

Hoepner, A. G. (2009). A categorisation of the responsible investment literature. *SSRN Working Paper 1283646.*

Hoepner, A. G., A. A. Majoch, and X. Y. Zhou (2019). Does an asset owner's institutional setting influence its decision to sign the principles for responsible investment? *Journal of Business Ethics*, 1–26.

Hoepner, A. G., I. Oikonomou, Z. Sautner, L. T. Starks, and X. Zhou (2021). ESG shareholder engagement and downside risk. *SSRN Working Paper 2874252.*

Hoepner, A. G. and L. Schopohl (2018). On the price of morals in markets: An empirical study of the Swedish AP-funds and the Norwegian government pension fund. *Journal of Business Ethics 151*(3), 665–692.

Hoepner, A. G. and P.-S. Yu (2017a). Corporate social responsibility across industries: when can who do well by doing good? *SSRN Working Paper 1284703.*

Hoepner, A. G. and P.-S. Yu (2017b). Responsible investors and company standards: Follow the money to rate the raters. *SSRN Working Paper 3023851.*

Hong, H. and M. Kacperczyk (2009). The price of sin: The effects of social norms on markets. *Journal of Financial Economics 93*(1), 15–36.

Hong, H. and L. Kostovetsky (2012). Red and blue investing: Values and finance. *Journal of Financial Economics 103*(1), 1–19.

Howarth, R. B. (2003). Discounting and uncertainty in climate change policy analysis. *Land Economics 79*(3), 369–381.

Hsu, A. W.-h. and T. Wang (2013). Does the market value corporate response to climate change? *Omega 41*(2), 195–206.

Hsu, P.-H., K. Li, and C.-Y. Tsou (2021). The pollution premium. *SSRN Working Paper 3578215*.

Hsu, S.-L. (2012). *The case for a carbon tax: Getting past our hang-ups to effective climate policy.* Island Press.

Huang, M., M. Li, X. Li, and M. Zhang (2021). Can star CEOs become morally better? The role of star CEOs on environmental, social and governance performance. *SSRN Working Paper 3897726*.

Huang, S. and G. Hilary (2018). Zombie board: board tenure and firm performance. *Journal of Accounting Research 56*(4), 1285–1329.

Hübel, B. and H. Scholz (2020). Integrating sustainability risks in asset management: the role of ESG exposures and ESG ratings. *Journal of Asset Management 21*(1), 52–69.

Hugon, A. and K. Law (2019). Impact of climate change on firm earnings: evidence from temperature anomalies. *SSRN Working Paper 3271386*.

Humphrey, J., S. Kogan, J. S. Sagi, and L. T. Starks (2021). The asymmetry in responsible investing preferences. *SSRN Working Paper 3583862*.

Humphrey, J. E., D. D. Lee, and Y. Shen (2012). Does it cost to be sustainable? *Journal of Corporate Finance 18*(3), 626–639.

Humphrey, J. E. and Y. Li (2021). Who goes green: Reducing mutual fund emissions and its consequences. *Journal of Banking & Finance Forthcoming*, 106098.

Humphrey, J. E. and D. T. Tan (2014). Does it really hurt to be responsible? *Journal of Business Ethics 122*(3), 375–386.

Hussaini, M., N. Hussain, D. K. Nguyen, and U. Rigoni (2021). Is corporate social responsibility an agency problem? an empirical note from takeovers. *Finance Research Letters Forthcoming*, 102007.

Husse, T. and F. Pippo (2021). Responsible minus irresponsible-a determinant of equity risk premia? *Journal of Sustainable Finance & Investment*, 1–23.

Huynh, T., F. W. Li, and Y. Xia (2021). Something in the air: Does air pollution affect fund managers' carbon divestment? *SSRN Working Paper 3908963*.

Hvidkjær, S. (2017). ESG investing: a literature review. *Working Paper*.

Ibikunle, G. and T. Steffen (2017). European green mutual fund performance: A comparative analysis with their conventional and black peers. *Journal of Business Ethics 145*(2), 337–355.

Ikram, A., Z. F. Li, and D. Minor (2019). CSR-contingent executive compensation contracts. *Journal of Banking & Finance*, 105655.

Ilhan, E., P. Krueger, Z. Sautner, and L. T. Starks (2021). Climate risk disclosure and institutional investors. *SSRN Working Paper 3437178*.

Iliev, P., J. Kalodimos, and M. Lowry (2021). Investors' attention to corporate governance. *Review of Financial Studies Forthcoming*.

In, S. Y., K. Y. Park, and R. G. Eccles (2020). What does carbon data tell you (or not)? *SSRN Working Paper 3754098*.

In, S. Y., K. Y. Park, and A. Monk (2017). Is 'being green'rewarded in the market? an empirical investigation of decarbonization risk and stock returns. *International Association for Energy Economics (Singapore Issue) 46*, 48.

In, S. Y. and K. Schumacher (2021). Carbonwashing- a new type of carbon data-related ESG greenwashing. *SSRN Working Paper 3901278*.

Indahl, R. and H. G. Jacobsen (2019). Private Equity 4.0: Using ESG to create more value with less risk. *Journal of Applied Corporate Finance 31*(2), 34–41.

Ioannou, I. and G. Serafeim (2015). The impact of corporate social responsibility on investment recommendations: Analysts' perceptions and shifting institutional logics. *Strategic Management Journal 36*(7), 1053–1081.

Ioannou, I. and G. Serafeim (2019). The consequences of mandatory corporate sustainability reporting. In *Oxford Handbook of Corporate Social Responsibility: Psychological and Organizational Perspectives*. Oxford University Press.

Jacobs, B. W., V. R. Singhal, and R. Subramanian (2010). An empirical investigation of environmental performance and the market value of the firm. *Journal of Operations Management 28*(5), 430–441.

Jacobsen, B., W. Lee, and C. Ma (2019). The alpha, beta, and sigma of ESG: Better beta, additional alpha? *Journal of Portfolio Management 45*(6), 6–15.

Jagannathan, R., A. Ravikumar, and M. Sammon (2018). Environmental, social, and governance criteria: Why investors should care. *Journal of Investment Management 16*(1), 18–31.

Jayachandran, S., K. Kalaignanam, and M. Eilert (2013). Product and environmental social performance: Varying effect on firm performance. *Strategic Management Journal 34*(10), 1255–1264.

Jha, A. and J. Cox (2015). Corporate social responsibility and social capital. *Journal of Banking & Finance 60*, 252–270.

Jia, J. and Z. Li (2021). Corporate sustainability, earnings persistence and the association between earnings and future cash flows. *Accounting & Finance Forthcoming*.

Jiang, R. and C. Weng (2020). Climate change risk and agriculture-related stocks. *SSRN Working Paper 3506311*.

Jiao, Y. (2010). Stakeholder welfare and firm value. *Journal of Banking & Finance 34*(10), 2549–2561.

Jin, I. (2020). ESG-screening and factor-risk-adjusted performance: the concentration level of screening does matter. *Journal of Sustainable Finance & Investment Forthcoming*, 1–21.

Jiraporn, P., N. Jiraporn, A. Boeprasert, and K. Chang (2014). Does corporate social responsibility (CSR) improve credit ratings? Evidence from geographic identification. *Financial Management 43*(3), 505–531.

Jo, H. and M. A. Harjoto (2011). Corporate governance and firm value: The impact of corporate social responsibility. *Journal of Business Ethics 103*(3), 351–383.

Jo, H. and H. Na (2012). Does CSR reduce firm risk? Evidence from controversial industry sectors. *Journal of Business Ethics 110*(4), 441–456.

Johnson, J. A., J. Theis, A. Vitalis, and D. Young (2020). The influence of firms' emissions management strategy disclosures on investors' valuation judgments. *Contemporary Accounting Research 37*(2), 642–664.

Johnson, R. A. and D. W. Greening (1999). The effects of corporate governance and institutional ownership types on corporate social performance. *Academy of Management Journal 42*(5), 564–576.

Johnson, S. A., T. C. Moorman, and S. Sorescu (2009). A reexamination of corporate governance and equity prices. *Review of Financial Studies 22*(11), 4753–4786.

Jørgensen, P. L. and M. D. Plovst (2021). The cost of insuring against underperformance in ESG screened index funds. *SSRN Working Paper 3831924*.

Jost, S., S. Erben, P. Ottenstein, and H. Zülch (2021). Does corporate social responsibility impact mergers & acquisition premia? New international evidence. *Finance Research Letters Forthcoming*, 102237.

Jung, J., K. Herbohn, and P. Clarkson (2018). Carbon risk, carbon risk awareness and the cost of debt financing. *Journal of Business Ethics 150*(4), 1151–1171.

Kacperczyk, M., J. Nosal, and L. Stevens (2019). Investor sophistication and capital income inequality. *Journal of Monetary Economics 107*, 18–31.

Kahn, M. E., K. Mohaddes, R. N. C Ng, M. H. Pesaran, M. Raissi, and J.-C. Yang (2019). Long-term macroeconomic effects of climate change: A cross-country analysis. *SSRN Working Paper 3428610*.

Kaiser, L. (2020a). Board effectiveness and firm risk. *Journal of Impact and ESG Investing 1*(2), 68–86.

Kaiser, L. (2020b). ESG integration: Value, growth and momentum. *Journal of Asset Management 21*(1), 32–51.

Kang, C., F. Germann, and R. Grewal (2016). Washing away your sins? Corporate social responsibility, corporate social irresponsibility, and firm performance. *Journal of Marketing 80*(2), 59–79.

Kang, J.-K., H. Kim, J. Kim, and A. Low (2021). Activist-appointed directors. *Journal of Financial and Quantitative Analysis Forthcoming*.

Kang, M., K. Viswanathan, N. A. White, and E. J. Zychowicz (2021). Sustainability efforts, index recognition, and stock performance. *Journal of Asset Management 22*, 120–132.

Kanuri, S. (2020). Risk and return characteristics of environmental, social, and governance (ESG) equity ETFs. *Journal of Index Investing 11*(2), 66–75.

Kaplan, R. S. and K. Ramanna (2021). How to fix ESG reporting. *SSRN Working Paper 3900146*.

Karakostas, A., G. Morgan, and D. J. Zizzo (2021). Socially interdependent investment. *SSRN Working Paper 3799854*.

Karwowski, M. and M. Raulinajtys-Grzybek (2021). The application of corporate social responsibility (CSR) actions for mitigation of environmental, social, corporate governance (ESG) and reputational risk in integrated reports. *Corporate Social Responsibility and Environmental Management 28*(4), 1270–1284.

Kempf, A. and P. Osthoff (2007). The effect of socially responsible investing on portfolio performance. *European Financial Management 13*(5), 908–922.

Khajenouri, D. C. and J. H. Schmidt (2020). Standard or sustainable-which offers better performance for the passive investor? *Journal of Applied Finance & Banking 11*(1), 61–71.

Khan, M. (2019). Corporate governance, ESG, and stock returns around the world. *Financial Analysts Journal 75*(4), 103–123.

Kiley, M. T. (2021). Growth at risk from climate change. *SSRN Working Paper 3907717*.

Kim, C.-S. (2019). Can socially responsible investments be compatible with financial performance? A meta-analysis. *Asia-Pacific Journal of Financial Studies 48*(1), 30–64.

Kim, E.-H. and T. P. Lyon (2015). Greenwash vs. brownwash: Exaggeration and undue modesty in corporate sustainability disclosure. *Organization Science 26*(3), 705–723.

Kim, H.-D., T. Kim, Y. Kim, and K. Park (2019). Do long-term institutional investors promote corporate social responsibility activities? *Journal of Banking & Finance 101*, 256–269.

Kim, I., J. W. Ryou, and R. Yang (2020). The color of shareholders' money: Institutional shareholders' political values and corporate environmental disclosure. *Journal of Corporate Finance 64*, 101704.

Kim, J.-B., B. Li, and Z. Liu (2018). Does social performance influence breadth of ownership? *Journal of Business Finance & Accounting 45*(9-10), 1164–1194.

Kim, S. and A. Yoon (2020). Analyzing active managers' commitment to ESG: Evidence from united nations principles for responsible investment. *SSRN Working Paper 3555984*.

Kim, T. and Y. Kim (2020). Capitalizing on sustainability: The value of going green. *SSRN Working Paper 3310643*.

King, A. and M. Lenox (2002). Exploring the locus of profitable pollution reduction. *Management Science 48*(2), 289–299.

King, A. A. and M. J. Lenox (2001). Does it really pay to be green? An empirical study of firm environmental and financial performance. *Journal of Industrial Ecology 5*(1), 105–116.

King, M., B. Tarbush, and A. Teytelboym (2019). Targeted carbon tax reforms. *European Economic Review 119*, 526–547.

Kitzmueller, M. and J. Shimshack (2012). Economic perspectives on corporate social responsibility. *Journal of Economic Literature 50*(1), 51–84.

Klassen, R. D. and C. P. McLaughlin (1996). The impact of environmental management on firm performance. *Management Science 42*(8), 1199–1214.

Klenert, D., L. Mattauch, E. Combet, O. Edenhofer, C. Hepburn, R. Rafaty, and N. Stern (2018). Making carbon pricing work for citizens. *Nature Climate Change 8*(8), 669–677.

Kling, G., U. Volz, V. Murinde, and S. Ayas (2021). The impact of climate vulnerability on firms' cost of capital and access to finance. *World Development 137*, 105131.

Knoll, M. S. (2002). Ethical screening in modern financial markets: the conflicting claims underlying socially responsible investment. *The Business Lawyer*, 681–726.

Kölbel, J. F., F. Heeb, F. Paetzold, and T. Busch (2020). Can sustainable investing save the world? Reviewing the mechanisms of investor impact. *Organization & Environment 33*(4), 554–574.

Konar, S. and M. A. Cohen (2001). Does the market value environmental performance? *Review of Economics and Statistics 83*(2), 281–289.

Kong, X., Y. Pan, H. Sun, and F. Taghizadeh-Hesary (2020). Can environmental corporate social responsibility reduce firms' idiosyncratic risk? evidence from China. *Frontiers in Environmental Science.*

Kordsachia, O., M. Focke, and P. Velte (2021). Do sustainable institutional investors contribute to firms' environmental performance? Empirical evidence from Europe. *Review of Managerial Science*, 1–28.

Kotsantonis, S., C. Pinney, and G. Serafeim (2016). ESG integration in investment management: Myths and realities. *Journal of Applied Corporate Finance 28*(2), 10–16.

Kotsantonis, S. and G. Serafeim (2019). Four things no one will tell you about ESG data. *Journal of Applied Corporate Finance 31*(2), 50–58.

Kreiser, L., M. S. Andersen, B. E. Olsen, S. Speck, and J. E. Milne (2015). *Carbon pricing: design, experiences and issues.* Edward Elgar Publishing.

Krosinsky, C. and N. Robins (2012). *Sustainable investing: The art of long-term performance.* Routledge.

Krosinsky, C., N. Robins, and S. Viederman (2011). *Evolutions in sustainable investing: strategies, funds and thought leadership*, Volume 618. John Wiley & Sons.

Krueger, P., Z. Sautner, and L. T. Starks (2020). The importance of climate risks for institutional investors. *Review of Financial Studies 33*(3), 1067–1111.

Krueger, P., Z. Sautner, D. Y. Tang, and R. Zhong (2021). The effects of mandatory ESG disclosure around the world. *SSRN Working Paper 3832745.*

Krüger, P. (2015). Corporate goodness and shareholder wealth. *Journal of Financial Economics 115*(2), 304–329.

Kuchler, T. and J. Stroebel (2020). Social finance. *SSRN Working Paper 3725297.*

Kumar, A., W. Xin, and C. Zhang (2019). Climate sensitivity and predictable returns. *SSRN Working Paper 3331872*.

Kuo, T.-C., H.-M. Chen, and H.-M. Meng (2021). Do corporate social responsibility practices improve financial performance? a case study of airline companies. *Journal of Cleaner Production 310*, 127380.

Lagerkvist, C., A. Edenbrandt, I. Tibbelin, and Y. Wahlstedt (2020). Preferences for sustainable and responsible equity funds: A choice experiment with Swedish private investors. *Journal of Behavioral and Experimental Finance 28*, 100406.

Lanfear, M. G., A. Lioui, and M. G. Siebert (2019). Market anomalies and disaster risk: Evidence from extreme weather events. *Journal of Financial Markets 46*, 100477.

Lanfear, M. G., A. Lioui, and M. G. Siebert (2020). Shelter from the storm: Which safe asset for climate disasters? *SSRN Working Paper 3511079*.

Lanoie, P., J. Laurent-Lucchetti, N. Johnstone, and S. Ambec (2011). Environmental policy, innovation and performance: new insights on the Porter hypothesis. *Journal of Economics & Management Strategy 20*(3), 803–842.

Lanoie, P., M. Patry, and R. Lajeunesse (2008). Environmental regulation and productivity: testing the Porter hypothesis. *Journal of Productivity Analysis 30*(2), 121–128.

Lanza, A., E. Bernardini, and I. Faiella (2020). Mind the gap! machine learning, ESG metrics and sustainable investment. *SSRN Working Paper 3659584*.

Lashitew, A. A. (2021). Corporate uptake of the Sustainable Development Goals: Mere greenwashing or an advent of institutional change? *Journal of International Business Policy 4*, 184–200.

Laufer, W. S. (2003). Social accountability and corporate greenwashing. *Journal of Business Ethics 43*(3), 253–261.

Le Treut, H. (2007). Historical overview of climate change. *Climate Change 2007: The Physical Science Basis. Contribution of Working Group I to the Fourth Assessment Report of the Intergovernmental Panel on Climate Change.*

Lee, D. D. and R. W. Faff (2009). Corporate sustainability performance and idiosyncratic risk: A global perspective. *Financial Review 44*(2), 213–237.

Lee, J. and E. Kim (2021). Would overconfident CEOs engage more in environment, social, and governance investments? With a focus on female representation on boards. *Sustainability 13*(6), 3373.

Lee, J., N. Pati, and J. J. Roh (2011). Relationship between corporate sustainability performance and tangible business performance: evidence from oil and gas industry. *International Journal of Business Insights and Transformation 3*(3), 72–82.

Lee, K.-H., B. Min, and K.-H. Yook (2015). The impacts of carbon (CO2) emissions and environmental research and development (R&D) investment on firm performance. *International Journal of Production Economics 167*, 1–11.

Lee, L.-E., G. Giese, and Z. Nagy (2020). Combining E, S, and G scores: An exploration of alternative weighting schemes. *Journal of Impact and ESG Investing 1*(1), 94–103.

Lemoine, D. and C. Traeger (2014). Watch your step: optimal policy in a tipping climate. *American Economic Journal: Economic Policy 6*(1), 137–66.

Lending, C., K. Minnick, and P. J. Schorno (2018). Corporate governance, social responsibility, and data breaches. *Financial Review 53*(2), 413–455.

Li, F. and A. Polychronopoulos (2020). What a difference an ESG ratings provider makes! *Research Affiliates Working Paper*.

Li, Q., H. Shan, Y. Tang, and V. Yao (2020). Corporate climate risk: Measurements and responses. *SSRN Working Paper 3508497*.

Li, Z., D. B. Minor, J. Wang, and C. Yu (2019). A learning curve of the market: Chasing alpha of socially responsible firms. *Journal of Economic Dynamics and Control 109*, 103772.

Li, Z. F. (2018). A survey of corporate social responsibility and corporate governance. In *Research Handbook of Finance and Sustainability*. Edward Elgar Publishing.

Li, Z. F., S. Patel, and S. Ramani (2020). The role of mutual funds in corporate social responsibility. *Journal of Business Ethics Forthcoming*, 1–23.

Li, Z. F. and C. Thibodeau (2019). CSR-contingent executive compensation incentive and earnings management. *Sustainability 11*(12), 3421.

Liang, H. and L. Renneboog (2018). Is corporate social responsibility an agency problem? In *Research Handbook of Finance and Sustainability*. Edward Elgar Publishing.

Liang, H. and L. Renneboog (2021). Corporate social responsibility and sustainable finance. In *Oxford Research Encyclopedia of Economics and Finance*.

Liang, H., L. Sun, and M. Teo (2021). Greenwashing: Evidence from hedge funds. *SSRN Working Paper 3610627*.

Lioui, A. (2018a). ESG factor investing: Myth or reality? *SSRN Working Paper 3272090*.

Lioui, A. (2018b). Is ESG risk priced? *SSRN Working Paper 3285091*.

Lioui, A., P. Poncet, and M. Sisto (2018). Corporate social responsibility and the cross section of stock returns. *SSRN Working Paper 2730722*.

Lioui, A. and Z. Sharma (2012). Environmental corporate social responsibility and financial performance: Disentangling direct and indirect effects. *Ecological Economics 78*, 100–111.

Lioui, A. and A. Tarelli (2021). Chasing the ESG factor. *SSRN Working Paper 3878314*.

Little, K. (2008). *The Complete Idiot's Guide to Socially Responsible Investing: Put Your Money Where Your Values Are*. Penguin.

Liu, B. and Y. Xu (2017). Is air-quality a risk factor that affects stock returns? *SSRN Working Paper 3103043*.

Liu, Z., Z. Deng, P. Ciais, R. Lei, S. J. Davis, S. Feng, B. Zheng, D. Cui, X. Dou, P. He, et al. (2020). Global daily CO2 emissions for the year 2020. *arXiv Preprint* (2103.02526).

Long, F. J. and S. Johnstone (2021). Applying 'deep ESG' to asian private equity. *Journal of Sustainable Finance & Investment Forthcoming*, 1–19.

Lopez, C., O. Contreras, and J. Bendix (2020a). Disagreement among ESG rating agencies: shall we be worried? *Munich Personal RePEc Archive Working Paper*.

Lopez, C., O. Contreras, and J. Bendix (2020b). ESG ratings: the road ahead. *SSRN Working Paper 3706440*.

López-Arceiz, F. J., A. J. Bellostas-Pérezgrueso, and J. M. Moneva (2018). Evaluation of the cultural environment's impact on the performance of the socially responsible investment funds. *Journal of Business Ethics 150*(1), 259–278.

Lopez de Silanes, F., J. A. McCahery, and P. Pudschedl (2020). ESG performance and disclosure: A cross-country analysis. *SSRN Working Paper 3506084.*

Lourenço, I. C., M. C. Branco, J. D. Curto, and T. Eugénio (2012). How does the market value corporate sustainability performance? *Journal of Business Ethics 108*(4), 417–428.

Lukomnik, J. and J. P. Hawley (2021). *Moving Beyond Modern Portfolio Theory: Investing that Matters.* Routledge.

Lund, A. C. and R. Schonlau (2016). Golden parachutes, severance, and firm value. *Fla. L. Rev. 68*, 875.

Luo, H. A. and R. J. Balvers (2017). Social screens and systematic investor boycott risk. *Journal of Financial and Quantitative Analysis 52*(1), 365–399.

Lyon, T. P., M. A. Delmas, J. W. Maxwell, P. Bansal, M. Chiroleu-Assouline, P. Crifo, R. Durand, J.-P. Gond, A. King, M. Lenox, et al. (2018). CSR needs CPR: Corporate sustainability and politics. *California Management Review 60*(4), 5–24.

Lyon, T. P. and A. W. Montgomery (2015). The means and end of greenwash. *Organization & Environment 28*(2), 223–249.

Ma, R., B. R. Marshall, H. Nguyen, N. H. Nguyen, and N. Visaltanachoti (2021). Climate events and return co-movement. *SSRN Working Paper 3898878.*

Madhavan, A., A. Sobczyk, and A. Ang (2021). Toward ESG alpha: Analyzing ESG exposures through a factor lens. *Financial Analysts Journal 77*(1), 69–88.

Madison, N. and E. Schiehll (2021). The effect of financial materiality on ESG performance assessment. *Sustainability 13*(7), 3652.

Mahmoud, O. (2020). Doing well while feeling good. *SSRN Working Paper 3458277*.

Mahmoud, O. and J. Meyer (2020). The anatomy of sustainability. *SSRN Working Paper 3597700*.

Mahmoud, O. and J. Meyer (2021). Morals, markets, and crises: Evidence from the covid pandemic. *SSRN Working Paper 3774995*.

Majoch, A. A., A. G. Hoepner, and T. Hebb (2017). Sources of stakeholder salience in the responsible investment movement: why do investors sign the principles for responsible investment? *Journal of Business Ethics 140*(4), 723–741.

Makridis, C. (2021). How extreme temperatures affect the formation of economic sentiment and its implications for stock returns. *SSRN Working Paper 3095422*.

Margolis, J. D., H. A. Elfenbein, and J. P. Walsh (2009). Does it pay to be good... and does it matter? a meta-analysis of the relationship between corporate social and financial performance. *SSRN Working Paper 1866371*.

Margot, V., C. Geissler, C. de Franco, B. Monnier, et al. (2021). ESG investments: Filtering versus machine learning approaches. *Applied Economics and Finance 8*(2), 1–16.

Mariani, M., F. Pizzutilo, A. Caragnano, and M. Zito (2021). Does it pay to be environmentally responsible? Investigating the effect on the weighted average cost of capital. *Corporate Social Responsibility and Environmental Management Forthcoming*.

Marquis, C., M. W. Toffel, and Y. Zhou (2016). Scrutiny, norms, and selective disclosure: A global study of greenwashing. *Organization Science 27*(2), 483–504.

Marshall, B. R., H. Nguyen, N. H. Nguyen, N. Visaltanachoti, and M. R. Young (2021). A note on green investment: Do climate disasters matter? *SSRN Working Paper 3788390*.

Marshall, B. R., N. H. Nguyen, and N. Visaltanachoti (2021). Financial markets and carbon dioxide emissions. *SSRN Working Paper 3884639*.

Martin, M. (2020). The contemporary history of impact investing. In *Social Aims of Finance*, pp. 85–110.

Martínez-Ferrero, J. and M.-B. Lozano (2021). The nonlinear relation between institutional ownership and Environmental, Social and Governance performance in emerging countries. *Sustainability 13*(3), 1586.

Martínez-Ferrero, J., L. Rodríguez-Ariza, and I.-M. García-Sánchez (2016). Corporate social responsibility as an entrenchment strategy, with a focus on the implications of family ownership. *Journal of Cleaner Production 135*, 760–770.

Matallín-Sáez, J. C., A. Soler-Domínguez, S. Navarro-Montoliu, and D. V. de Mingo-López (2021). Investor behavior and the demand for conventional and socially responsible mutual funds. *Corporate Social Responsibility and Environmental Management*.

Matallín-Sáez, J. C., A. Soler-Domínguez, E. Tortosa-Ausina, and D. V. de Mingo-López (2019). Ethical strategy focus and mutual fund management: Performance and persistence. *Journal of Cleaner Production 213*, 618–633.

Matos, P. (2020). *ESG and Responsible Institutional Investing Around the World: A Critical Review*. CFA Institute Research Foundation.

Matos, P. V., V. Barros, and J. M. Sarmento (2020). Does ESG affect the stability of dividend policies in Europe? *Sustainability 12*(21), 1–15.

Matsumura, E. M., R. Prakash, and S. C. Vera-Munoz (2014). Firm-value effects of carbon emissions and carbon disclosures. *Accounting Review 89*(2), 695–724.

Matsumura, E. M., R. Prakash, and S. C. Vera-Muñoz (2020). Climate-risk materiality and firm risk. *SSRN Working Paper 2983977*.

Mayer, C. (2019). Valuing the invaluable: how much is the planet worth? *Oxford Review of Economic Policy 35*(1), 109–119.

McCahery, J. A., Z. Sautner, and L. T. Starks (2016). Behind the scenes: The corporate governance preferences of institutional investors. *Journal of Finance 71*(6), 2905–2932.

McGuire, J. B., A. Sundgren, and T. Schneeweis (1988). Corporate social responsibility and firm financial performance. *Academy of Management Journal 31*(4), 854–872.

McKibbin, W. J., A. C. Morris, P. J. Wilcoxen, and A. J. Panton (2020). Climate change and monetary policy: issues for policy design and modelling. *Oxford Review of Economic Policy 36*(3), 579–603.

McWilliams, A. and D. Siegel (2000). Corporate social responsibility and financial performance: correlation or misspecification? *Strategic Management Journal 21*(5), 603–609.

McWilliams, A. and D. Siegel (2001). Corporate social responsibility: A theory of the firm perspective. *Academy of Management Review 26*(1), 117–127.

Mehran, H. (1995). Executive compensation structure, ownership, and firm performance. *Journal of Financial Economics 38*(2), 163–184.

Mervelskemper, L. (2018). How ESG information determines emotional returns of socially responsible investments: Evidence from an experimental decision case. *SSRN Working Paper 3107751*.

Mervelskemper, L. and D. Streit (2017). Enhancing market valuation of ESG performance: Is integrated reporting keeping its promise? *Business Strategy and the Environment 26*(4), 536–549.

Metcalf, G. E. (2018). *Paying for pollution: why a carbon tax is good for America.* Oxford University Press.

Meuer, J., J. Koelbel, and V. H. Hoffmann (2020). On the nature of corporate sustainability. *Organization & Environment 33*(3), 319–341.

Michalski, L. and R. K. Y. Low (2021). Corporate credit rating feature importance: Does ESG matter? *SSRN Working Paper 3788037.*

Michelon, G. and A. Parbonetti (2012). The effect of corporate governance on sustainability disclosure. *Journal of Management & Governance 16*(3), 477–509.

Milne, J. E. and M. S. Andersen (2012). *Handbook of research on environmental taxation.* Edward Elgar Publishing.

Mittnik, S., W. Semmler, and A. Haider (2020). Climate disaster risks—empirics and a multi-phase dynamic model. *Econometrics 8*(3), 1–27.

Mizuno, T., S. Doi, T. Tsuchiya, and S. Kurizaki (2021). Socially responsible investing through the equity funds in the global ownership network. *Plos One 16*(8), e0256160.

Mohamed Buallay, A., M. Al Marri, N. Nasrallah, A. Hamdan, E. Barone, and Q. Zureigat (2021). Sustainability reporting in banking and financial services sector: a regional analysis. *Journal of Sustainable Finance & Investment Forthcoming*, 1–26.

Mohanty, S. S., O. Mohanty, and M. Ivanof (2021). Alpha enhancement in global equity markets with ESG overlay on factor-based investment strategies. *Risk Management 23*, 213–242.

Monasterolo, I. (2020). Climate change and the financial system. *Annual Review of Resource Economics 12*, 299–320.

Monasterolo, I. and L. De Angelis (2020). Blind to carbon risk? an analysis of stock market reaction to the paris agreement. *Ecological Economics 170*, 106571.

Monti, A., P. Pattitoni, B. Petracci, and O. Randl (2019). Does corporate social responsibility impact equity risk? International evidence. *SSRN Working Paper 3167883*.

Mooij, S. (2017a). Asset managers' ESG strategy: Lifting the veil. *SSRN Working Paper 3123218*.

Mooij, S. (2017b). The ESG rating and ranking industry; vice or virtue in the adoption of responsible investment? *SSRN Working Paper 2960869*.

Morgan, J. and J. Tumlinson (2019). Corporate provision of public goods. *Management Science 65*(10), 4489–4504.

Morgenstern, C., G. Coqueret, and J. Kelly (2021). Tuning trend following strategies with macro ESG data. *Journal of Impact and ESG Investing Forthcoming*.

Moss, A., J. P. Naughton, and C. Wang (2020). The irrelevance of ESG disclosure to retail investors: Evidence from robinhood. *SSRN Working Paper 3604847*.

Moussa, T., A. Allam, S. Elbanna, and A. Bani-Mustafa (2020). Can board environmental orientation improve US firms' carbon performance? The mediating role of carbon strategy. *Business Strategy and the Environment 29*(1), 72–86.

Muñoz, F. (2021a). Carbon-intensive industries in socially responsible mutual funds' portfolios. *International Review of Financial Analysis Forthcoming*, 101740.

Muñoz, F. (2021b). Sin sectors and negative screening. *Business Ethics, the Environment & Responsibility 30*(2), 216–230.

Murata, R. and S. Hamori (2021). ESG disclosures and stock price crash risk. *Journal of Risk and Financial Management 14*(2), 70.

Mutlu, C. C., M. Van Essen, M. W. Peng, S. F. Saleh, and P. Duran (2018). Corporate governance in China: A meta-analysis. *Journal of Management Studies 55*(6), 943–979.

Naaraayanan, S. L., K. Sachdeva, and V. Sharma (2021). The real effects of environmental activist investing. *SSRN Working Paper 3483692*.

Nadaraja, S., A. Huang, B. Liu, and S. Ali (2020). Does board gender diversity reduce default risk? A global analysis. In *Academy of Management Proceedings*. Academy of Management Briarcliff Manor, NY 10510.

Naffa, H. and M. Fain (2020). Performance measurement of ESG-themed megatrend investments in global equity markets using pure factor portfolios methodology. *PloS One 15*(12).

Naffa, H. and M. Fain (2021). A factor approach to the performance of ESG leaders and laggards. *Finance Research Letters Forthcoming*, 102073.

Nagy, Z., A. Kassam, and L.-E. Lee (2016). Can ESG add alpha? An analysis of ESG tilt and momentum strategies. *Journal of Investing 25*(2), 113–124.

Naranjo Tuesta, Y., C. Crespo Soler, and V. Ripoll Feliu (2020). The influence of carbon management on the financial performance of European companies. *Sustainability 12*(12), 4951.

Narassimhan, E., K. S. Gallagher, S. Koester, and J. R. Alejo (2018). Carbon pricing in practice: A review of existing emissions trading systems. *Climate Policy 18*(8), 967–991.

Nath, S. (2021). The business of virtue: Evidence from socially responsible investing in financial markets. *Journal of Business Ethics 169*, 1811–199.

Nekhili, M., H. Nagati, T. Chtioui, and C. Rebolledo (2017). Corporate social responsibility disclosure and market value: Family versus nonfamily firms. *Journal of Business Research 77*, 41–52.

Neubaum, D. O. and S. A. Zahra (2006). Institutional ownership and corporate social performance: The moderating effects of investment horizon, activism, and coordination. *Journal of Management 32*(1), 108–131.

Newell, R. G., W. A. Pizer, and B. C. Prest (2021). A discounting rule for the social cost of carbon.

Ng, A. and D. Zheng (2018). Let's agree to disagree! On payoffs and green tastes in green energy investments. *Energy Economics 69*, 155–169.

Ng, A. C. and Z. Rezaee (2015). Business sustainability performance and cost of equity capital. *Journal of Corporate Finance 34*, 128–149.

Ng, A. C. and Z. Rezaee (2020). Business sustainability factors and stock price informativeness. *Journal of Corporate Finance 64*, 101688.

Nofsinger, J. and A. Varma (2014). Socially responsible funds and market crises. *Journal of Banking & Finance 48*, 180–193.

Nofsinger, J. R., J. Sulaeman, and A. Varma (2019). Institutional investors and corporate social responsibility. *Journal of Corporate Finance 58*, 700–725.

Noh, D. and S. Oh (2021). Measuring institutional pressure for greenness: A demand system approach. *SSRN Working Paper 3639693*.

Nugent, T., N. Stelea, and J. L. Leidner (2020). Detecting ESG topics using domain-specific language models and data augmentation approaches. *arXiv Preprint* (2010.08319).

Nwachukwu, C. (2021). Systematic review of integrated reporting: recent trend and future research agenda. *Journal of Financial Reporting and Accounting Forthcoming*.

Oikonomou, I., C. Brooks, and S. Pavelin (2012). The impact of corporate social performance on financial risk and utility: A longitudinal analysis. *Financial Management 41*(2), 483–515.

Oikonomou, I., E. Platanakis, and C. Sutcliffe (2018). Socially responsible investment portfolios: Does the optimization process matter? *British Accounting Review 50*(4), 379–401.

Okafor, A., M. Adusei, and B. N. Adeleye (2021). Corporate social responsibility and financial performance: Evidence from US tech firms. *Journal of Cleaner Production*, 126078.

Omura, A., E. Roca, and M. Nakai (2020). Does responsible investing pay during economic downturns: Evidence from the COVID-19 pandemic. *Finance Research Letters 42*, 101914.

Ooi, S. K., A. Amran, J. A. Yeap, and A. H. Jaaffar (2019). Governing climate change: the impact of board attributes on climate change disclosure. *International Journal of Environment and Sustainable Development 18*(3), 270–288.

Orlitzky, M., F. L. Schmidt, and S. L. Rynes (2003). Corporate social and financial performance: A meta-analysis. *Organization Studies 24*(3), 403–441.

Ouchen, A. (2021). Is the ESG portfolio less turbulent than a market benchmark portfolio? *Risk Management*, 1–33.

Palazzolo, C., L. Pomorski, and A. Zhao (2020). (car) bon voyage: The road to low carbon investment portfolios. *SSRN Working Paper 3753023*.

Park, J. h. and J. H. Noh (2018). Relationship between climate change risk and cost of capital. *Global Business & Finance Review 23*(2), 66–81.

Parker, F. J. (2021). Achieving goals while making an impact: Balancing financial goals with impact investing. *Journal of Impact and ESG Investing Forthcoming*.

Partnership for Market Readiness, a. (2017). *Carbon tax guide: a handbook for policy makers*. World Bank.

Pastor, L., R. F. Stambaugh, and L. A. Taylor (2021a). Dissecting green returns. *SSRN Working Paper 3864502*.

Pastor, L., R. F. Stambaugh, and L. A. Taylor (2021b). Sustainable investing in equilibrium. *Journal of Financial Economics 142*(2), 550–571.

Pástor, L. and M. B. Vorsatz (2020). Mutual fund performance and flows during the COVID-19 crisis. *Review of Asset Pricing Studies 10*(4), 791–833.

Patel, P. C., J. A. Pearce II, and P. Oghazi (2020). Not so myopic: Investors lowering short-term growth expectations under high industry ESG-sales-related dynamism and predictability. *Journal of Business Research Forthcoming*.

Patuelli, R., P. Nijkamp, and E. Pels (2005). Environmental tax reform and the double dividend: A meta-analytical performance assessment. *Ecological economics 55*(4), 564–583.

Pavlova, I. and M. E. de Boyrie (2021). ESG ETFs and the COVID-19 stock market crash of 2020: Did clean funds fare better? *Finance Research Letters Forthcoming*, 102051.

Pedersen, L. H., S. Fitzgibbons, and L. Pomorski (2021). Responsible investing: The ESG-efficient frontier. *Journal of Financial Economics 142*(2), 572–597.

Peiris, D. and J. Evans (2010). The relationship between environmental social governance factors and US stock performance. *Journal of Investing 19*(3), 104–112.

Pelizzon, L., A. Rzeznik, and K. Weiss-Hanley (2021). The salience of ESG ratings for stock pricing: Evidence from (potentially) confused investors.

Pindyck, R. S. (2013). Climate change policy: What do the models tell us? *Journal of Economic Literature 51*(3), 860–72.

Piu, S. (2020). ESG investing: What does the research say? *MAN Institute Research Paper*.

Pizzutilo, F. (2021). Is ESG-ness the vaccine? *Applied Economics Letters Forthcoming*.

Plagge, J.-C. and D. M. Grim (2020). Have investors paid a performance price? Examining the behavior of ESG equity funds. *Journal of Portfolio Management 46*(3), 123–140.

Porter, M. E. and C. Van der Linde (1995). Toward a new conception of the environment-competitiveness relationship. *Journal of Economic Perspectives 9*(4), 97–118.

Post, C. and K. Byron (2015). Women on boards and firm financial performance: A meta-analysis. *Academy of Management Journal 58*(5), 1546–1571.

Price, J. M. and W. Sun (2017). Doing good and doing bad: The impact of corporate social responsibility and irresponsibility on firm performance. *Journal of Business Research 80*, 82–97.

Pronobis, P. and F. Venuti (2021). Accounting for sustainability: Current initiatives to standardize ESG reporting. *ESCP Working Paper*.

Pyles, M. K. (2020). Examining portfolios created by bloomberg ESG scores: Is disclosure an alpha factor? *Journal of Impact and ESG Investing 1*(2), 39–52.

Radhouane, I., M. Nekhili, H. Nagati, and G. Paché (2018). The impact of corporate environmental reporting on customer-related performance and market value. *Management Decision 57*(7), 1630–1659.

Rafaty, R., G. Dolphin, and F. Pretis (2021). Carbon pricing and the elasticity of CO_2 emissions. *SSRN Working Paper 3812786*.

Raimo, N., A. Caragnano, M. Zito, F. Vitolla, and M. Mariani (2021). Extending the benefits of ESG disclosure: The effect on the cost of debt financing. *Corporate Social Responsibility and Environmental Management 28*(4), 1412–1421.

Raman, N., G. Bang, and A. Nourbakhsh (2020). Mapping ESG trends by distant supervision of neural language models. *Machine Learning and Knowledge Extraction 2*(4), 453–468.

Ramelli, S., E. Ossola, and M. Rancan (2020). Climate sin stocks: Stock price reactions to global climate strikes. *SSRN Working Paper 3544669*.

Ramelli, S., A. F. Wagner, R. J. Zeckhauser, and A. Ziegler (2021). Investor rewards to climate responsibility: Stock-price responses to the opposite shocks of the 2016 and 2020 US elections. *Review of Corporate Financial Studies Forthcoming*.

Ramsey, F. P. (1928). A mathematical theory of saving. *Economic Journal 38*(152), 543–559.

Rankin, C. P. (2020). Under pressure: exploring pressures for corporate social responsibility in mutual funds. *Qualitative Research in Organizations and Management: An International Journal*.

Rathner, S. (2013). The influence of primary study characteristics on the performance differential between socially responsible and conventional investment funds: A meta-analysis. *Journal of Business Ethics 118*(2), 349–363.

Ravina, A. and R. Hentati Kaffel (2020). The impact of low-carbon policy on stock returns. *SSRN Working Paper 3444168.*

Raynaud, J., P. Tankov, and S. Voisin (2020). Portfolio alignment to a 2°C trajectory: Science or art? *SSRN Working Paper 3644171.*

Reber, B., A. Gold, and S. Gold (2021). ESG disclosure and idiosyncratic risk in initial public offerings. *Journal of Business Ethics Forthcoming*, 1–20.

Reinders, H. J., D. Schoenmaker, and M. A. Van Dijk (2020). A finance approach to climate stress testing. *SSRN Working Paper 3594231.*

Renneboog, L., J. Ter Horst, and C. Zhang (2008a). The price of ethics and stakeholder governance: The performance of socially responsible mutual funds. *Journal of Corporate Finance 14*(3), 302–322.

Renneboog, L., J. Ter Horst, and C. Zhang (2008b). Socially responsible investments: Institutional aspects, performance, and investor behavior. *Journal of Banking & Finance 32*(9), 1723–1742.

Renneboog, L., J. Ter Horst, and C. Zhang (2011). Is ethical money financially smart? nonfinancial attributes and money flows of socially responsible investment funds. *Journal of Financial Intermediation 20*(4), 562–588.

Revelli, C. and J.-L. Viviani (2015). Financial performance of socially responsible investing (SRI): what have we learned? A meta-analysis. *Business Ethics: A European Review 24*(2), 158–185.

Reynolds, N.-S. and M. E. Hassett (2021). *The role of sustainability agency in mergers and acquisitions*, pp. 248–258. Edward Elgar Publishing.

Rezaee, Z. and L. Tuo (2019). Are the quantity and quality of sustainability disclosures associated with the innate and discretionary earnings quality? *Journal of Business Ethics 155*(3), 763–786.

Rhodes, M. J. (2010). Information asymmetry and socially responsible investment. *Journal of Business Ethics 95*(1), 145–150.

Richey, G. M. (2020). Is it good to sin when times are bad? An investigation of the defensive nature of sin stocks. *Journal of Investing 29*(6), 43–50.

Riedl, A. and P. Smeets (2017). Why do investors hold socially responsible mutual funds? *Journal of Finance 72*(6), 2505–2550.

Rissman, P. and D. Kearney (2019). Rise of the shadow ESG regulators: Investment advisers, sustainability accounting, and their effects on corporate social responsibility. *Envtl. L. Rep. News & Analysis 49*, 10155.

Roncalli, T., T. L. Guenedal, F. Lepetit, T. Roncalli, and T. Sekine (2020). Measuring and managing carbon risk in investment portfolios. *arXiv Preprint* (2008.13198).

Roncalli, T., T. L. Guenedal, F. Lepetit, T. Roncalli, and T. Sekine (2021). The market measure of carbon risk and its impact on the minimum variance portfolio. *Journal of Portfolio Management Forthcoming.*

Roselle, P. (2016). The evolution of integrating ESG analysis into wealth management decisions. *Journal of Applied Corporate Finance 28*(2), 75–79.

Rossi, M., J. Chouaibi, S. Chouaibi, W. Jilani, and Y. Chouaibi (2021). Does a board characteristic moderate the relationship between CSR practices and financial performance? Evidence from European ESG firms. *Journal of Risk and Financial Management 14*(8), 354.

Rossi, M., D. Sansone, A. Van Soest, and C. Torricelli (2019). Household preferences for socially responsible investments. *Journal of Banking & Finance 105*, 107–120.

Roulet, T. J. and S. Touboul (2015). The intentions with which the road is paved: Attitudes to liberalism as determinants of greenwashing. *Journal of Business Ethics 128*(2), 305–320.

Roundy, P., H. Holzhauer, and Y. Dai (2017). Finance or philanthropy? Exploring the motivations and criteria of impact investors. *Social Responsibility Journal 13*(3), 491–512.

Russo, M. V. and P. A. Fouts (1997). A resource-based perspective on corporate environmental performance and profitability. *Academy of Management Journal 40*(3), 534–559.

Sachs, J., W. T. Woo, N. Yoshino, and F. Taghizadeh-Hesary (2019). *Handbook of Green Finance: Energy Security and Sustainable Development.* Springer.

Saeed, A., U. Noreen, A. Azam, and M. S. Tahir (2021). Does CSR governance improve social sustainability and reduce the carbon footprint: International evidence from the energy sector. *Sustainability 13*(7), 3596.

Sahin, Ö., K. Bax, S. Paterlini, and C. Czado (2021). ESGM: ESG scores and the missing pillar. *SSRN Working Paper* (3890696).

Santi, C. (2021). Investors' climate sentiment and financial markets. *SSRN Working Paper 3697581*.

Sautner, Z. and L. T. Starks (2021). ESG and downside risks: implications for pension funds. *SSRN Working Paper 3879170*.

Sautner, Z., L. van Lent, G. Vilkov, and R. Zhang (2021a). Firm-level climate change exposure. *SSRN Working Paper 3642508*.

Sautner, Z., L. van Lent, G. Vilkov, and R. Zhang (2021b). Pricing climate change exposure. *SSRN Working Paper 3792366*.

Scanlan, M. K. (2021). Climate risk is investment risk. *Journal of Environmental Law & Litigation 36*.

Schanzenbach, M. M. and R. H. Sitkoff (2020). ESG investing: Theory, evidence, and fiduciary principles. *Journal of Financial Planning Forthcoming*.

Schlenker, W. and C. A. Taylor (2021). Market expectations of a warming climate. *Journal of Financial Economics 142*(2), 627–640.

Schmidt, A. B. (2020). Optimal ESG portfolios: an example for the dow jones index. *Journal of Sustainable Finance & Investment*, 1–7.

Schmidt, A. B. and X. Zhang (2021). Optimal ESG portfolios: Which ESG ratings to use? *SSRN Working Paper 3859674*.

Schoenmaker, D. and W. Schramade (2018). *Principles of Sustainable Finance*. Oxford University Press.

Schoenmaker, D. and W. Schramade (2019). Investing for long-term value creation. *Journal of Sustainable Finance & Investment 9*(4), 356–377.

Schröder, M. (2004). The performance of socially responsible investments: investment funds and indices. *Financial Markets and Portfolio Management 18*(2), 122–142.

Schueth, S. (2003). Socially responsible investing in the United States. *Journal of Business Ethics 43*(3), 189–194.

Seele, P. and L. Gatti (2017). Greenwashing revisited: In search of a typology and accusation-based definition incorporating legitimacy strategies. *Business Strategy and the Environment 26*(2), 239–252.

Semenova, N. and L. G. Hassel (2008). Financial outcomes of environmental risk and opportunity for US companies. *Sustainable Development 16*(3), 195–212.

Semieniuk, G., E. Campiglio, J.-F. Mercure, U. Volz, and N. R. Edwards (2021). Low-carbon transition risks for finance. *Wiley Interdisciplinary Reviews: Climate Change 12*(1), e678.

Sen, S. and M.-T. von Schickfus (2020). Climate policy, stranded assets, and investors' expectations. *Journal of Environmental Economics and Management 100*, 102277.

Serafeim, G. (2020). Public sentiment and the price of corporate sustainability. *Financial Analysts Journal 76*(2), 26–46.

Serafeim, G. and J. Grewal (2017). The value relevance of corporate sustainability disclosures: An analysis of a dataset from one large asset owner. *SSRN Working Paper 2966767*.

Serafeim, G. and A. Yoon (2021a). Stock price reactions to ESG news: The role of ESG ratings and disagreement. *Review of Accounting Studies Forthcoming*.

Serafeim, G. and A. Yoon (2021b). Which corporate ESG news does the market react to? *Financial Analysts Journal Forthcoming*.

Servaes, H. and A. Tamayo (2013). The impact of corporate social responsibility on firm value: The role of customer awareness. *Management Science 59*(5), 1045–1061.

Shafer, M. and E. Szado (2020). Environmental, social, and governance practices and perceived tail risk. *Accounting & Finance 60*(4), 4195–4224.

Shanaev, S. and B. Ghimire (2021). When ESG meets AAA: The effect of ESG rating changes on stock returns. *Finance Research Letters Forthcoming*, 102302.

Sharfman, B. S. (2020). ESG investing under ERISA. *JREG Bulletin 38*, 112.

Sharfman, M. P. and C. S. Fernando (2008). Environmental risk management and the cost of capital. *Strategic Management Journal 29*(6), 569–592.

Sharma, G. D., G. Talan, S. Bansal, and M. Jain (2021). Is there a cost for sustainable investments: evidence from dynamic conditional correlation. *Journal of Sustainable Finance & Investment*, 1–21.

Shaw, C., S. G. Evans, and J. Turner (2021). Evaluating economic output at risk to climate change: A sectoral comparison of economic sensitivity to weather. *Extreme Events and Climate Change: A Multidisciplinary Approach*, 205–217.

Shen, S., A. LaPlante, and A. Rubtsov (2019). Strategic asset allocation with climate change. *SSRN Working Paper 3249211*.

Sherwood, M. W. and J. Pollard (2018). *Responsible investing: An introduction to environmental, social, and governance investments*. Routledge.

Shive, S. A. and M. M. Forster (2020). Corporate governance and pollution externalities of public and private firms. *Review of Financial Studies 33*(3), 1296–1330.

Siano, A., A. Vollero, F. Conte, and S. Amabile (2017). "more than words": Expanding the taxonomy of greenwashing after the Volkswagen scandal. *Journal of Business Research 71*, 27–37.

Silvola, H. (2021). *Sustainable Investing: Beating the Market with ESG*. Springer Nature.

Simerly, R. L. (1995). Institutional ownership, corporate social performance, and firms' financial performance. *Psychological Reports 77*(2), 515–525.

Simpson, W. G. and T. Kohers (2002). The link between corporate social and financial performance: Evidence from the banking industry. *Journal of Business Ethics 35*(2), 97–109.

Singh, A. (2020). COVID-19 and safer investment bets. *Finance Research Letters 36*, 101729.

Singh, A. (2021). COVID-19 and ESG preferences: Corporate bonds versus equities. *International Review of Finance Forthcoming*.

Singhania, M. and N. Saini (2021). Institutional framework of ESG disclosures: comparative analysis of developed and developing countries. *Journal of Sustainable Finance & Investment Forthcoming*, 1–44.

Soana, M.-G. (2011). The relationship between corporate social performance and corporate financial performance in the banking sector. *Journal of Business Ethics 104*(1), 133.

Sokolov, A., K. Caverly, J. Mostovoy, T. Fahoum, and L. Seco (2021). Weak supervision and Black-Litterman for automated ESG portfolio construction. *Journal of Financial Data Science Forthcoming*.

Sokolov, A., J. Mostovoy, J. Ding, and L. Seco (2021). Building machine learning systems for automated ESG scoring. *Journal of Impact and ESG Investing Forthcoming*.

Starks, L. T., P. Venkat, and Q. Zhu (2020). Corporate ESG profiles and investor horizons. *SSRN Working Paper 3049943*.

Statman, M. (2000). Socially responsible mutual funds (corrected). *Financial Analysts Journal 56*(3), 30–39.

Statman, M. (2006). Socially responsible indexes. *Journal of Portfolio Management 32*(3), 100–109.

Statman, M. (2020). ESG as waving banners and as pulling plows. *Journal of Portfolio Management 46*(3), 16–25.

Statman, M. and D. Glushkov (2009). The wages of social responsibility. *Financial Analysts Journal 65*(4), 33–46.

Statman, M. and D. Glushkov (2016). Classifying and measuring the performance of socially responsible mutual funds. *Journal of Portfolio Management 42*(2), 140–151.

Stephenson, D. and V. Vracheva (2015). Corporate social responsibility and tax avoidance: A literature review and directions for future research. *SSRN Working Paper 2756640*.

Stips, A., D. Macias, C. Coughlan, E. Garcia-Gorriz, and X. San Liang (2016). On the causal structure between CO_2 and global temperature. *Scientific Reports 6*(1), 1–9.

Stotz, O. (2021). Expected and realized returns on stocks with high-and low-ESG exposure. *Journal of Asset Management 22*, 133–150.

Stroebel, J. and J. Wurgler (2021). What do you think about climate finance? *Journal of Financial Economics 142*(2), 487–498.

Stubbs, W. and P. Rogers (2013). Lifting the veil on environment-social-governance rating methods. *Social Responsibility Journal 9*(4), 622–640.

Stultz, R. S. (2020). An examination of the efficacy of christian-based socially responsible mutual funds. *Journal of Impact and ESG Investing 1*(2), 105–119.

Symitsi, E. and P. Stamolampros (2020). Employee sentiment and stock return predictability. *SSRN Working Paper 3640661*.

Talan, G. and G. D. Sharma (2019). Doing well by doing good: A systematic review and research agenda for sustainable investment. *Sustainability 11*(2), 353.

Taleb, W., T. Le Guenedal, F. Lepetit, V. Mortier, T. Sekine, and L. Stagnol (2020). Corporate ESG news and the stock market. *SSRN Working Paper 3723799*.

Tang, D. Y., J. Yan, and C. Y. Yao (2021). The determinants of ESG ratings: Rater ownership matters. *SSRN Working Paper 3889395*.

Tankov, P. and A. Tantet (2019). Climate data for physical risk assessment in finance. *SSRN Working Paper 3480156*.

Tasnia, M., S. M. S. J. AlHabshi, and R. Rosman (2021). The impact of corporate social responsibility on stock price volatility of the US banks: a moderating role of tax. *Journal of Financial Reporting and Accounting 19*(1), 77–91.

Taticchi, P. and M. Demartini (2020). *Corporate Sustainability in Practice*. Springer.

Terzani, S. and T. Turzo (2021). Religious social norms and corporate sustainability: The effect of religiosity on environmental, social, and governance disclosure. *Corporate Social Responsibility and Environmental Management 28*, 485–496.

Timár, B. et al. (2021). How does the market price responsible and sustainable investments? *Financial and Economic Review 20*(2), 117–147.

Tirodkar, M. (2020). Is pollution a sin? A study on the institutional ownership of polluter stocks. *SSRN Working Paper 3085411*.

Tokat-Acikel, Y., M. Aiolfi, L. D. Johnson, J. Hall, and Y. Jin (2021). Top-down portfolio implications of climate change. *Journal of Portfolio Management Forthcoming*.

Tol, R. S. (2019). *Climate economics: economic analysis of climate, climate change and climate policy*. Edward Elgar Publishing.

Tol, R. S. (2021). The economic impact of weather and climate. *SSRN Working Paper* (3807009).

Torre, M. L., F. Mango, A. Cafaro, and S. Leo (2020). Does the ESG index affect stock return? Evidence from the Eurostoxx 50. *Sustainability 12*(16), 6387.

Townsend, B. (2020). From SRI to ESG: The origins of socially responsible and sustainable investing. *Journal of Impact and ESG Investing 1*(1), 10–25.

Traeger, C. P. (2014). Why uncertainty matters: discounting under intertemporal risk aversion and ambiguity. *Economic Theory 56*(3), 627–664.

Traeger, C. P. (2021). Uncertainty in the analytic climate economy. *SSRN Working Paper 3846154*.

Trinks, P. J. and B. Scholtens (2017). The opportunity cost of negative screening in socially responsible investing. *Journal of Business Ethics 140*(2), 193–208.

Tsai, H.-J. and Y. Wu (2021). Changes in corporate social responsibility and stock performance. *Journal of Business Ethics Forthcoming*, 1–21.

Tsukioka, Y. (2018). Does the mandatory adoption of outside directors improve firm performance? *SSRN Working Paper 3220311*.

Umar, Z., D. Kenourgios, and S. Papathanasiou (2020). The static and dynamic connectedness of environmental, social, and governance investments: International evidence. *Economic Modelling 93*, 112–124.

Utz, S. (2019). Corporate scandals and the reliability of ESG assessments: evidence from an international sample. *Review of Managerial Science 13*(2), 483–511.

Van Beurden, P. and T. Gössling (2008). The worth of values–A literature review on the relation between corporate social and financial performance. *Journal of Business Ethics 82*(2), 407.

Van Duuren, E., A. Plantinga, and B. Scholtens (2016). ESG integration and the investment management process: Fundamental investing reinvented. *Journal of Business Ethics 138*(3), 525–533.

Van Nes, E. H., M. Scheffer, V. Brovkin, T. M. Lenton, H. Ye, E. Deyle, and G. Sugihara (2015). Causal feedbacks in climate change. *Nature Climate Change 5*(5), 445–448.

Varmaz, A., C. Fieberg, and T. Poddig (2021). Portfolio optimization for sustainable investments. *SSRN Working Paper 3859616*.

Velte, P. (2016). Women on management board and ESG performance. *Journal of Global Responsibility 7*(1), 98–109.

Velte, P. (2017). Does ESG performance have an impact on financial performance? Evidence from Germany. *Journal of Global Responsibility*.

Velte, P. (2019). The bidirectional relationship between ESG performance and earnings management–empirical evidence from Germany. *Journal of Global Responsibility 10*(4).

Venmans, F., J. Ellis, and D. Nachtigall (2020). Carbon pricing and competitiveness: are they at odds? *Climate Policy 20*(9), 1070–1091.

Verheyden, T., R. G. Eccles, and A. Feiner (2016). ESG for all? The impact of ESG screening on return, risk, and diversification. *Journal of Applied Corporate Finance 28*(2), 47–55.

Vojtko, R. and M. Padysak (2019). Quant's look on ESG investing strategies. *SSRN Working Paper 3504767.*

Von Arx, U. and A. Ziegler (2014). The effect of corporate social responsibility on stock performance: New evidence for the USA and Europe. *Quantitative Finance 14*(6), 977–991.

Waddock, S. A. and S. B. Graves (1997). The corporate social performance–financial performance link. *Strategic Management Journal 18*(4), 303–319.

Wagemans, F. A., C. K. v. Koppen, and A. P. Mol (2013). The effectiveness of socially responsible investment: a review. *Journal of Integrative Environmental Sciences 10*(3-4), 235–252.

Walker, K. and F. Wan (2012). The harm of symbolic actions and green-washing: Corporate actions and communications on environmental performance and their financial implications. *Journal of Business Ethics 109*(2), 227–242.

Walter, I. (2020). Sense and nonsense in ESG ratings. *Journal of Law, Finance and Accounting 5*(2), 307–336.

Wee, K.-W., Y.-S. Kang, J. M. Chung, and J. Lee (2020). The effect of ESG levels on fund performance and cash flows. *Financial Planning Review 13*(2), 83–115.

Weitzman, M. L. (2009). On modeling and interpreting the economics of catastrophic climate change. *Review of Economics and Statistics 91*(1), 1–19.

Wendt, V.-S., J. Kazdin, K. Schwaiger, and A. Ang (2021). Climate alpha with predictors also improving firm efficiency. *SSRN Working Paper 3889640*.

Wernicke, G., M. Sajko, and C. Boone (2021). How much influence do ceos have on company actions and outcomes? the example of corporate social responsibility. *Academy of Management Discoveries Forthcoming*.

Whelan, T., U. Atz, T. Van Holt, and C. Clark (2021). ESG and financial performance: Uncovering the relationship by aggregating evidence from 1,000 plus studies published between 2015 – 2020. Technical report, NYU Stern, Center for Sustainable Business.

Widyawati, L. (2020). A systematic literature review of socially responsible investment and environmental social governance metrics. *Business Strategy and the Environment 29*(2), 619–637.

Windsor, D. (2013). Corporate social responsibility and irresponsibility: A positive theory approach. *Journal of Business Research 66*(10), 1937–1944.

Wu, G.-z. and D. You (2021). " stabilizer" or" catalyst"? How does green technology innovation affect the risk of stock price crash: an analysis based on the quantity and quality of patents. *arXiv Preprint* (2106.16177).

Wu, Q. and J. Lu (2020). Air pollution, individual investors, and stock pricing in china. *International Review of Economics & Finance 67*, 267–287.

Xiao, Y., R. Faff, P. Gharghori, and B.-K. Min (2017). The financial performance of socially responsible investments: Insights from the intertemporal CAPM. *Journal of Business Ethics 146*(2), 353–364.

Xie, J., W. Nozawa, M. Yagi, H. Fujii, and S. Managi (2019). Do environmental, social, and governance activities improve corporate financial performance? *Business Strategy and the Environment 28*(2), 286–300.

Xiong, J. X. (2021). The impact of ESG risk on stocks. *Journal of Impact and ESG Investing Forthcoming.*

Xu, H., J. Wu, and M. Dao (2020). Corporate social responsibility and trade credit. *Review of Quantitative Finance and Accounting 54*, 1389–1416.

Xu, J., C. Sun, and Y. You (2021). Climate change exposure and stock return predictability. *SSRN Working Paper 3777060.*

Yan, S., F. Ferraro, and J. Almandoz (2019). The rise of socially responsible investment funds: The paradoxical role of the financial logic. *Administrative Science Quarterly 64*(2), 466–501.

Yang, R. (2021). What do we learn from ratings about corporate social responsibility (CSR)? *SSRN Working Paper 3165783.*

Yang, Z., T. T. H. Nguyen, H. N. Nguyen, T. T. N. Nguyen, and T. T. Cao (2020). Greenwashing behaviours: causes, taxonomy and consequences based on a systematic literature review. *Journal of Business Economics and Management 21*(5), 1486–1507.

Yao, S., Y. Pan, A. Sensoy, G. S. Uddin, and F. Cheng (2021). Green credit policy and firm performance: What we learn from China. *Energy Economics 101*, 105415.

Yen, T.-Y. and P. André (2019). Market reaction to the effect of corporate social responsibility on mergers and acquisitions: Evidence on emerging markets. *Quarterly Review of Economics and Finance 71*, 114–131.

Yin, H., M. Li, Y. Ma, and Q. Zhang (2019). The relationship between environmental information disclosure and profitability: A comparison between different disclosure styles.

International Journal of Environmental Research and Public Health 16(9), 1556.

Yoo, S. and S. Managiy (2021). Disclosure or action: Evaluating ESG behavior towards financial performance. *Finance Research Letters Forthcoming*, 102108.

Yousaf, I., T. Suleman, and R. Demirer (2021). Green investments: A luxury good or a financial necessity? *SSRN Working Paper 3855125.*

Yu, E. P.-y. and B. Van Luu (2021). International variations in ESG disclosure - Do cross-listed companies care more? *International Review of Financial Analysis 75*, 101731.

Zaccone, M. C. and M. Pedrini (2020). ESG factor integration into private equity. *Sustainability 12*(14), 5725.

Zeng, J. and G. Strobl (2016). The effect of activists' short-termism on corporate governance. *SSRN Working Paper 2736052.*

Zhan, J. X. and A. U. Santos-Paulino (2021). Investing in the sustainable development goals: Mobilization, channeling, and impact. *Journal of International Business Policy 4*, 166–183.

Zhang, X., X. Zhao, and L. Qu (2021). Do green policies catalyze green investment? Evidence from ESG investing developments in China. *Economics Letters 207*, 110028.

Zhao, C. L. and P. P. Tans (2006). Estimating uncertainty of the WMO mole fraction scale for carbon dioxide in air. *Journal of Geophysical Research: Atmospheres 111*(D8).

Zhao, X. and A. J. Murrell (2016). Revisiting the corporate social performance-financial performance link: A replication of Waddock and Graves. *Strategic Management Journal 37*(11), 2378–2388.

Ziegler, A., M. Schröder, and K. Rennings (2007). The effect of environmental and social performance on the stock performance of European corporations. *Environmental and Resource Economics 37*(4), 661–680.

Ziolo, M. (2020). *Finance and Sustainable Development: Designing Sustainable Financial Systems.* Routledge.

Zumente, I. and J. Bistrova (2021). ESG importance for long-term shareholder value creation: Literature vs. practice. *Journal of Open Innovation: Technology, Market, and Complexity 7*(2), 127.

Index

B-Corp, 25
board, 21, 26, 41, 53, 55, 58,
    65, 69
bonds, 67
boycott, 59
brownwashing, 24

carbon constraints, 77
carbon emissions, 27
carbon neutral portfolio, 82
carbon pricing, 99
carbon tax, 25, 35, 76, 98, 99
carbon-washing, 25
causality, 19, 55, 91
central banks, 100
CEO, 34, 40
climate change, 85
climate disasters, 94–97
climate risk, 86
CO_2 emissions, 41, 58, 61,
    89–91
communication, 23, 52
corporate environmental
    performance, 5
corporate financial
    performance, 5
corporate scandals, 22
corporate social
    responsibility, 3
cost of equity, 69
COVID-19, 28, 34, 66, 90

customers, 23, 34, 42, 57

disagreement, 20
disclosure, 9, 12, 14, 15
discounting, 86

earnings, 28, 34, 42, 53, 69,
    94
efficient frontier, 59, 79
employee satisfaction, 52, 69
environmental Kuznets
    curve, 18
ESG funds, 8
ESG incidents, 42, 55, 68
ESG issues, 12
ESG momentum, 52

fake ESG investing, 44
footprint, 15, 23, 24, 27, 31,
    37, 44, 58, 81, 83

gender parity, 19, 20, 38, 39,
    58
global warming, 85
governance, 26, 40, 41, 51
green firms, 25
greenhouse gas emissions, 15,
    41, 59, 89
greenwashing, 22–24, 65

impact, 38, 42
impact investing, 8, 38

incentives, 23, 25, 33, 34,
    38–41, 57, 98
institutional investors, 31,
    34, 36, 37, 40, 43, 64
investors, 31

machine learning, 95
missing data, 17
missing pillar, 17
monetary policy, 100

natural language processing,
    94
non-linearity, 64
nudging, 37

ownership, 39, 41, 42, 64

Paris agreement, 83, 95
performance attribution, 81
phyisical risk, 93
physical risk, 86, 92
politics, 35
Porter hypothesis, 98
portfolio optimization, 59,
    76, 78, 80
prediction, 90
private equity, 36
probability of default, 93

rating addiction, 30
rating disagreement, 19
rating heterogeneity, 21
rating provider, 11, 19
regulatory perspectives, 9
regulatory risk, 86
religion, 36
remuneration, 43

reporting, 11, 12, 25, 28, 29,
    37, 90, 96
retail custumers, 42
retail investors, 37, 43
return on equity, 69
risk, 21, 35, 37, 43, 45, 52,
    57, 66, 72, 85
risk management, 43

scope 3 emissions, 16
sentiment, 93
shareholder activism, 38
shareholder engagement, 40
social cost of carbon, 88
social discounting rate, 88
social pressure, 57
sovereign ESG, 18
stress test, 92
suppliers, 42
sustainable trap, 59

temperature, 96
text processing, 22
Tobin's q, 69
transition risk, 86, 93

uncertainty, 86, 99

venture capital, 37

warm glow, 33

Printed in the United States
by Baker & Taylor Publisher Services